Morell Mackenzie

Essay on Growths in the Larynx

With Reports and an Analysis or One Hundred Consecutive Cases Treated by the

Author

Morell Mackenzie

Essay on Growths in the Larynx
With Reports and an Analysis or One Hundred Consecutive Cases Treated by the Author

ISBN/EAN: 9783744785600

Printed in Europe, USA, Canada, Australia, Japan

Cover: Foto ©berggeist007 / pixelio.de

More available books at **www.hansebooks.com**

ESSAY ON

GROWTHS IN THE LARYNX:

WITH REPORTS, AND AN ANALYSIS OF

ONE HUNDRED CONSECUTIVE CASES TREATED BY THE AUTHOR,

AND

A TABULAR STATEMENT OF ALL PUBLISHED CASES TREATED

BY OTHER PRACTITIONERS

SINCE THE INVENTION OF THE LARYNGOSCOPE.

BY

MORELL MACKENZIE, M.D. Lond., M.R.C.P.

PHYSICIAN TO THE HOSPITAL FOR DISEASES OF THE THROAT, AND TO THE ROYAL SOCIETY
OF MUSICIANS; AND SENIOR ASSISTANT-PHYSICIAN, AND LECTURER ON
DISEASES OF THE THROAT, AT THE LONDON HOSPITAL.

WITH NUMEROUS ILLUSTRATIONS

IN CHROMO-LITHOGRAPHY AND WOOD-ENGRAVING.

PHILADELPHIA:

LINDSAY & BLAKISTON.

1871.

LONDON :

WYMAN AND SONS, PRINTERS, GREAT QUEEN STREET

LINCOLN'S-INN FIELDS, W.C.

TO

PROFESSOR CZERMAK, M.D.

BY WHOSE GENIUS AND PERSEVERANCE THE LARYNGOSCOPE

WAS BROUGHT TO PERFECTION ;

BY WHOM A GROWTH IN THE LARYNX WAS FIRST SEEN WITH THE

LARYNGEAL MIRROR ;

AND

WITHOUT WHOSE LABOURS THIS VOLUME COULD NOT

HAVE BEEN WRITTEN :

THESE PAGES ARE DEDICATED

BY HIS OBLIGED FRIEND AND PUPIL

THE AUTHOR.

PREFACE.

IN the year 1863, the Jacksonian Prize was awarded to me for my "Essay on Diseases of the Larynx." It was my intention to publish that work at the time, but my increasing engagements prevented me from at once carrying out this project, and Laryngology made such rapid advances that the views of one year became almost an anachronism in the next. Under these circumstances, I determined to publish, from time to time, a series of monographs on Diseases of the Throat, based on my original Prize Essay, but brought up to the most recent date.

For the subject of my first Essay I selected Nervo-Muscular Affections of the Larynx, on account of their very frequent occurrence and great importance; and the great interest which attaches to Laryngeal Growths has induced me to choose this subject for my second Essay.

Since Ehrmann published his classical *Histoire des Polypes du Larynx*, in the year 1850, the Laryngoscope has effected an entire change in our views, and it is thought that sufficient time has now elapsed to enable us to consider these diseases from the new standpoint.

The present Essay is based on an experience of nearly 150 cases of laryngeal growth. It includes detailed Reports of 112 cases, of which 26 have been previously published in the Medical journals or the Transactions of Medical Societies, and 86 are now brought forward for the first time.

Of my 112 cases, 100 underwent treatment, and 12 were merely observed with the laryngoscope.

The 100 cases are consecutive, no treated case having been intentionally omitted : 77 of the cases were cured, in 18 improvement took place, in 3 the result was negative, and 2 died. Of the fatal cases, one was an infant, and the other an adult. In both tracheotomy had been performed.

The circumstances of nearly every case, which I have successfully treated, are known to one or more medical practitioners, and in by far the greater number of cases, the growths have been removed in the presence of professional *confrères.*

I have excluded all cases of carcinoma, although in some instances great temporary benefit has resulted from the more or less complete removal of such neoplasms ; and I have also excluded all cases of " false excrescence."[1]

I have likewise omitted a few other cases,—some because the notes were too imperfect to be of any value, some because they were only seen in consultation with other practitioners, and one because (no treatment having been advised by me) the case was subsequently reported by another physician.

[1] See page 36.

There still remain a considerable number of true growths which have come under my care since June, 1870 ; but as for statistical purposes it was convenient to limit the number of cases treated to 100—a number probably sufficient to eliminate accidental circumstances—I have omitted all those subsequently seen.

In addition to my own cases, I have appended a record of all the published cases which have been treated since the invention of the Laryngoscope : all sources available in this country down to the end of the year 1870 have been carefully searched. In this arduous work, which relates to cases not only in England and America, but to a much greater number in Germany, France, Russia, Italy, Austria, and Hungary, I have received great assistance from Dr. JAGIELSKI and Dr. BAÜMLER.

My thanks are due to my colleagues at the London Hospital, Dr. ANDREW CLARK and Dr. FENWICK, for their kind assistance in the microscopic examination of many of my specimens, and to my brother, Mr. STEPHEN MACKENZIE, who has made careful microscopic drawings of the neoplasms removed by me during the last two years.

I am also greatly indebted to my assistant, Mr. LENNOX BROWNE, for much valuable help in bringing out this work. All the laryngoscopic drawings made for me since the year 1867 have been taken from life by him with great fidelity and artistic skill, and he has made finished drawings of those which I myself had sketched at an earlier date. In addition to this, he has greatly aided me by taking notes of the greater number of my cases during the last four years. My former pupil and friend, Mr. PUGIN THORNTON, has also given me valuable assistance by keeping records of many of my cases.

Lastly, I have to thank the numerous medical friends who have brought so many of the patients under my notice, and those who, by their presence at my public demonstrations, have encouraged me to persevere in the treatment of these laborious and difficult cases.

13, WEYMOUTH STREET, PORTLAND PLACE. W.,
June, 1871.

TABLE OF CONTENTS.

c

SECTION IV.

SECTION V.

SECTION VI.

SECTION VII.

APPENDIX A.

APPENDIX B.

APPENDIX C.

APPENDIX D.

ESSAY ON
GROWTHS IN THE LARYNX.

—◦◦—

BENIGN GROWTHS IN THE LARYNX.

DEFINITION :—

New formations of benign character, forming projections on the mucous membrane of the larynx, generally giving rise to aphonia or dysphonia, often to dyspnœa, and occasionally to dysphagia.

SYNONYMS :—

Latin.—Polypus Laryngis ; Tumor Laryngis.

French.—Polype du Larynx ; Tumeur Polypeuse ; Tumeur Verruqueuse ; Tumeur Fibreuse ; Excroissance Epithéliale ; Végétations Papillaires ; Kyste Muqueuse.

German. — Larynxpolyp ; Fibröspolyp ; Bindgewebsgeschwulst ; Papillärgeschwulst ; Schleimpolyp ; Neubildung.

English.—Polypus of the Larynx ; Fibrous Tumour ; Fibrous Growth ; Connective-tissue Growth ; Wart ; Warty Growth ; Excrescence ; Vegetation ; Cystic Tumour ; etc.

I

SECTION I.

HISTORY.

Isolated cases of laryngeal polypus are to be found from an early date, and the case in which Koderik successfully operated on a growth through the mouth, about the year 1750,[1] is one of the first described.

Seventeen years later, Lieutaud[2] published two cases of undoubted laryngeal polypus.

In the beginning of the present century the Imperial Government of France offered a prize for the best work on Croup, and several eminent practitioners in different parts of Europe turned their attention to diseases of the larynx.

---

[1] George Herbinaux, *Parallèle des différens Instruments, avec les Méthodes de s'en servir pour pratiquer la Ligature des Polpyes dans la Matrice, en forme de Lettre à M. Roux, avec Figures*. A la Haye, chez Gosse et Perel. 1771.

This case is quoted by Lewin in his able and comprehensive article in the *Deutsche Klinik*, March 29, 1862.

[2] "63. In cadavere cujusdam *asthmatici* triginta annorum, qui perpetuo querebatur de quodam impedimento in tracheâ, quod *tussi* et screatu expellere sæpius conabatur et morte subitanea sublati, reperitur quidam *polypus* variis radicibus laryngi infixus et versus glottidem obturamenti instar adactus, unde suffocatio inexpectata.

"64. Secto cadavere cujusdam pueri duodecim annorum jampridem *phthisici* et inexpectata morte rapti, in propatulum veniebat intra laryngem corpus quoddam *polyposum* et racemosum tracheæ superiori parte, pediculo unico et peculiari ortum trahens et hinc fluitans ; quo forte ad laryngem repulso suffocationem obierat æger."--Lieutaud, *Historia Anatom. med.*, lib. iv. observ. 63, 64. 1767.

In reporting these cases, Ehrmann, in his classical *Histoire des Polypes du Larynx* (Strasbourg, 1850), remarks at page 5 as follows :—" Les deux observa-

We hence find a number of cases observed at this time, and subsequently published. Albers, of Bonn, one of the successful competitors, published an important work[1] in the year 1833; and in the same year, Brauers, of Louvain, attempted to extirpate a laryngeal polypus by division of the thyroid cartilage. In 1836 Regnoli[2] recorded a case in which he removed a growth from the larynx through the mouth, after performing tracheotomy; and the following year Ryland[3] devoted several pages of his classical work to tumours of the larynx. It was not, however, until the year 1850 that a complete monograph appeared. Then it was that Ehrmann published his celebrated treatise,[4] which included thirty-one cases of laryngeal growth. In the year 1851[5] Rokitansky brought forward ten additional cases; and in 1852 Dr. Horace Green,[6] of New York, published thirty-nine cases, two of which had occurred in his own practice. In the following year, Dr. Gurdon Buck collected forty-nine cases, including his own interesting example; and in 1854 Middeldorpf[7] brought together sixty-four cases. Finally, in the year 1859, Prat published a case in which he had removed a growth through the thyro-hyoid membrane.[8]

Amongst all these cases, there are only nine in which an attempt was made to remove the growth during life, and one of these, viz. that by Koderik, already referred to, is so vague, that it must necessarily be excluded. Of the remaining eight cases, in four 'instances (those of Brauers, Ehrmann,

tions de Lieutaud paraissent fort concluantes; on voit que dans la première le mot *asthmatici* est souligné, ainsi que le mot *phthisici* dans la seconde; n'est-il pas probable que par cette simple marque typographique l'auteur ait voulu indiquer que les polypes avaient simulé, pendant la vie, les symptômes d'asthme et de phthisie qui en avaient imposé aux médecins?"

[1] *Dissert. de Tumorib. in Cavo Laryngis.* Bonn.
[2] *Osservazion. chirurg., &c.* Pisa, 1836.
[3] *A Treatise on the Diseases and Injuries of the Larynx and Trachea.*
[4] *Op. cit.*
[5] *Zeitschrift der k. k. Gesellschaft der Aerzte zu Wien*, März, 1851.
[6] *Polypi of the Larynx and Œdema of the Glottis.* New York, 1852. *Transactions of the American Medical Association*, 1853.
[7] *Die Galvanokaustik.* Breslau, 1854.
[8] *Gazette des Hôpitaux*, 1859, No. 103, p. 809.

Gurdon Buck, and Prat), direct incision was made through
the neck; whilst only in the cases of Regnoli, Professor
Middeldorpf, and the two cases of Dr. Horace Green, was
the growth more or less completely removed *per vias naturales.*   In Regnoli's case the patient was a peasant, aged
70, who suffered from great dyspnœa and dysphagia.   On
opening the mouth widely, a tumour, about the size of a
hen's egg, could be seen at the back of the throat, and, on
introducing the finger, its peduncle could be traced down
to the arytenoid cartilages.   After performing tracheotomy,
Dr. Regnoli removed the growth through the mouth.   Unfortunately, the growth returned, and had to be removed a
second time.   The patient soon after sank from weakness.[1]

Dr. Horace Green, as already remarked, reported two
cases : in one, he removed a pedunculated tumour about
the size of a cherry, which was (thought to be) attached to
the left vocal cord (but which may have grown from the
ventricular band).   When the mouth was widely opened, and
the patient coughed, a round white fibrous-looking tumour
could be seen projecting upwards between the ary-epiglottic
folds.   Green was able to seize the growth with the ordinary
tonsil forceps, and then divided it with a long slender knife.
Dr. Green's second case was that of a gentleman, aged 42,
who was suffering from aphonia, dyspnœa, and dysphagia.
By passing a sponge probang saturated with a strong solution
of nitrate of silver into the larynx, Dr. Green succeeded in
separating a number of small polypoid excrescences, varying in size from a millet-seed to a duck-shot.   The number
of particles dislodged at different operations exceeded thirty.
The breathing improved, but the aphonia remained.

In Professor Middeldorpf's[2] case he succeeded in removing
a tumour from the upper opening of the larynx, by means

---

[1] I have been unable to find Regnoli's work in this country, but I am informed
by my friend Dr. Massei, of Naples, who, at my request, kindly forwarded me
details of the case, that it is also reported in Professor Andrea Ranzi's edition of
Regnoli's *Lectures on Surgical Pathology,* vol. iv. Florence, 1850, as well as in
Osserv. chir. &c., 1836.

[2] *Of. cit.,* p. 212, and Ruhle : *Die Kehlkopfkrankheiten.*   Berlin, 1861, p. 229.

of the galvano-caustic wire. "The sarcomatous growth showed a high degree of cell-development," and as a portion remained behind, a very doubtful prognosis was given : solutions of nitrate of silver were afterwards used. Rühle, who saw the case six years after the operation, states "that there was no symptom at that time of any return of the growth." In addition to these recorded cases, there remains another, which was not published at the time. The patient was treated by Sir Astley Cooper,[1] who removed with his finger a large cancerous tumour, about the size of a hen's egg, from the under surface of the epiglottis. It grew again, and was again removed, and the patient finally died from hæmorrhage. This case is very similar in its clinical history to that of Regnoli, except that tracheotomy does not appear to have been necessary.

In perusing the various cases collected by different authors in ante-laryngoscopic times, it will be seen, that, in many instances, the growth was of distinctly carcinomatous character, and that others are altogether of a different kind, to those contained in this treatise.

The invention of the laryngoscope has naturally given a great impetus to the study of throat diseases, and since the genius and perseverance of Czermak placed the laryngeal mirror in the hands of the medical profession, a very great number of cases have been reported. To him, who brought the laryngoscope to perfection, was also reserved the honour of first discovering a growth with the laryngeal mirror. On the 2nd of January, 1859, Dr. Hirschler requested Dr. Czermak to see a patient, who for some years had been treated for "nervous hoarseness." On laryngoscopic examination, it was at once seen that the dysphonia was due to a small warty growth on the right vocal cord.[2]

Since that time, upwards of 200 cases have been published, in which laryngeal growths have been *treated* with

---

[1] The specimen is preserved in the Museum of Guy's Hospital (No. 1685). This case was first published by me, in the year 1865, in the first edition of my work on *The Use of the Laryngoscope*, p. 111 et seq.

[2] *Wien. med. Wochenschrift*, 8 Januar, 1859.

the aid of the mirror. When we compare the cases observed with the laryngoscope with the specimens found in pathological museums, we are struck with the enormous antagonism, as to the comparative frequency of laryngeal growths. In the museums of the Royal College of Surgeons and the various London Hospitals, there exist altogether only thirty-four specimens of true laryngeal growth, whilst in my own practice I have seen with the laryngoscope, in the course of ten years, more than four times as many cases. Dr. Krishaber[1] remarks that laryngeal growths " are to be met with in two or three per cent. of cases of disease of the larynx, exclusively local and chronic." In referring to this passage, Mr. Durham[2] observes that his own experience would lead him to "the conclusion that they are much less frequent even than this." I am unable to say how frequent laryngeal growths are, in comparison with other chronic local affections of the larynx, but I find, on an approximative analysis, that in relation to all other throat affections, including those of the pharynx, these cases have occurred in my private practice in the proportion of $1\frac{1}{2}$ per cent., whilst they have been present in only one-half per cent. of the cases under my care at the Hospital for Diseases of the Throat, and of my throat-cases at the London Hospital. The larger percentage occurring in private practice is not due to a greater liability of the upper and middle classes to this affection, but to the accidental circumstance, that the most common symptom—loss of voice—is of more consequence to the educated and wealthy than to those engaged in manual labour.

The great discrepancy between ante-laryngoscopic observations and present experience is to be accounted for by the following circumstances :—*First,* In former times, autopsies were, and even at present are, frequently made without the larynx being opened at all. *Secondly,* Even if the larynx be opened, a small vegetation may be easily overlooked. As has

---

[1] *Dict. Encyclopéd. des Sciences médicales,* Paris, 1868, art. "Laryngoscope."
[2] Holmes's *System of Surgery,* 2nd edition, vol. iv., art. "Diseases of the Larynx," p. 574.

been remarked by Lewin, it is extremely probable, that, in former times, many cases of sudden death, attributed to apoplexy, or heart disease, were due to spasm of the glottis caused by growths. This author well observes, that "We have now the opportunity of diagnosing many growths, which are so small, and are so situated, that though they give rise to symptoms, they do not cause death. If death subsequently take place from other causes, the larynx is either not opened at all, or is examined so superficially, that small growths are very likely not to be recognized. This is all the more likely to happen in consequence of the great diminution in volume which takes place in laryngeal growths after death."

Dr. Lewin also well remarks, that obstruction of the larynx may give rise to pulmonary apoplexy or œdema, and that these secondary affections have probably often been regarded as the true cause of death. To the objection, that growths in the larynx would have given rise to some laryngeal symptoms during life, and that an inspection of the windpipe would therefore have been made after death, he urges, that, in the greater number of cases recorded before the invention of the laryngoscope, either no diagnosis was made, or the symptoms were attributed to phthisis, asthma, croup, ulceration of the larynx, spasm of the glottis, &c.[1]

[1] *Deutsche Klinik*, 1862. Lewin's arguments have been slightly epitomized. In each of the diseases referred to as having been wrongly diagnosed, and where growths were susbequently found, he quotes numerous cases from Andral, Senn, Pelletan, Dupuytren, &c.

CAUSES.

Influence of Chronic Hyperæmia of the Mucous Membrane.— Chronic congestion of the mucous membrane of the larynx is, far above all other causes, the most important ætiological feature, in the production of simple morbid growths in the larynx. In some cases, the disease appears to originate in an acute or subacute form of inflammation, but it is generally only as the starting-point of chronic hyperæmia, that the more acute attack indirectly leads to the production of a new formation.

Prolonged irritation of the mucous membrane of the larynx was long ago referred to by Ehrmann and Horace Green as the determining cause of polypus of the larynx, and of course that "irritation" acts by giving rise to hyperæmia.

The most common cause of hyperæmia is probably catarrh, and catarrh must therefore be looked upon as the great predisponent of growths. The various other influences, hereafter considered, probably only act through establishing a condition of hyperæmia.

Influence of Dyscrasiæ.—Originating as the morbid affection does in local irritation, it is not surprising to find that diathesis appears to have little influence in its production. Neither syphilis nor phthisis, nor any other constitutional condition, appears to favour the growth of these neoplasms. In the later stages of laryngeal phthisis, imperfect papillary growths do occasionally appear on the posterior wall of the larynx, and on the mucous membrane covering the vocal

cords[1] and the inner surface of the arytenoid cartilages, but this is the exception. Indeed, both the diathetic conditions referred to, appear to exercise a decidedly antagonistic influence to the development of new formations. Seeing that growths most commonly originate in hyperæmia, and that both syphilis and phthisis frequently give rise to that condition, it might have been expected that these constitutional states would have often been present in cases of laryngeal growth. The fact, however, is, that the congestion of laryngeal phthisis is soon followed by submucous changes, which interfere with growth at the free surface ; and though, in the early forms of syphilis, the occasional presence of condylomata shows a tendency to the formation in excess of an imperfectly organized tissue, these manifestations are forms of eruption, which tend to spontaneous subsidence. When a very protracted syphilitic congestion occurs, growths may arise, but this is a rare exception ; and Dr. Harlan has well pointed out, that few laryngeal growths can be attributed to syphilis.[2] Of course laryngeal growths may occur in syphilitic persons as they do in the healthy, but syphilis does not appear to be a factor in their production. In the records of the specimens in the various Pathological Museums in London, syphilis is stated to have previously existed in several instances, and in a specimen in the London Hospital, the patient was the subject of hereditary syphilis. On the subject of dyscrasia, Virchow,[3] who includes not only true fibrous growths, but connective-tissue tumours and papillomata, in his comprehensive division of Fibromata, remarks as follows :—
" The various forms of fibroma are all essentially hyperplastic, completely homologous, indeed, or, as is otherwise said, hypertrophic; even in those cases, however, where numerous growths occur in the same situation, as for instance, on the skin, they cannot be attributed to a special

[1] In the case of a patient who recently (January, 1871) died from laryngeal phthisis, in the Hospital for Diseases of the Throat, a warty growth of considerable size was found on the left vocal cord. The neoplasm consisted entirely of very large epithelial cells.

[2] *American Journal of Medical Science*, vol. lii. p. 122.

[3] *Die krankhaften Geschwülste*, vol. i. p. 356.

fibromatous dyscrasia . . . . It often happens that on
the serous surfaces, a considerable elevation is found, consisting of larger and smaller fibrous nodules. These are
universally regarded as the expression of an inflammatory
irritation, which affected the whole surface, though it did
not give rise to the same result in all parts . . . . .
These cases prove that, within a given area of dermal or
connective tissue, a condition of vulnerability may exist,
which under slight influences may give rise to isolated
eruptions. The nodular tuberous fibromata are in this respect
closely parallel with the warty growths, which are very often
multiple, and which grow not only on the hands, in great
numbers, but are also found covering a large surface of the
rest of the body. Without further evidence, it cannot be
concluded that there is such a thing as a warty dyscrasia,
or a warty constitution of the body in general, for the skin is
the only part affected.   If we examine these cases, however,
more closely, we find that with regard to the hands, the
essential cause is external irritation, and that those persons,
who are not subjected to any irritation, suffer little or not at
all, from warts.   There is no doubt, that the class of working
people, who are not required to do much with their hands
suffer very little from warts, whilst cooks, coachmen, and
artisans are often very much troubled with them.   This is
evidently the result of their occupations, and other parts of
the body, under certain circumstances, show a similar predisposition. . . . . The greatest difference is seen, however,
to exist : one individual, under certain circumstances, being
affected with warts, whilst another, under the same conditions, is perfectly free ; in the same individual also, warts
are more readily formed at particular times than at others.
This proves that a variable predisposition exists in the
tissues, which are the seat of these growths.   This must
depend on the condition of the part, inasmuch as different
results are produced by the same external irritation.   If the
irritation which takes place be very transient, it may pass off
without any result.   Under these circumstances, I believe,
that, after all, we must fall back on local predisposition, which

also thoroughly explains the multiplicity of these tumours. If the views of humoral pathology be logically followed out, we are driven to the conclusion that warts are contagious (a popular view, which is still largely entertained), and that the contagion lies in the blood.

" It is an old superstition that, if, when a wart is cut, blood flow on the skin, a fresh wart arises : this follows naturally from the old humoral pathology, and, to be logical, must be based on the supposition that the blood in the wart is the essential carrier of contagion.

" The occurrence of those numerous cases of congenital fibroma, which increase from very slight irritation, and develop into large tumours, favours the idea of local predisposition. The much more rare hereditary forms, which develop after birth, and the multiplication of which is always limited to a single system, favour the same idea. The same may also be said of those numerous cases, in which slight wounds are the attributed causes of the development of fibromata, or where the essential tumour-formation takes place in a tissue, predisposed through previous morbid processes, as is so clearly seen in elephantiasis. I have repeatedly remarked, that I do not exclude the idea of a dyscrasia,—nay, that I even do not deny the specific nature of such a dyscrasia. Syphilis is here the best example, and both the acuminated and the broad condylomata have been attributed to this disease. . . . . . Should, however, a dyscrasic condition be the cause of the irritation, a fibroma is essentially a formation of local character, and in common language, benign. If indeed it grows, and spreads itself, it has little inclination to ulcerate. On the other hand, many of the tumours now brought under consideration, especially warts and condylomata, and even the slighter forms of elephantiasis, frequently disappear spontaneously, being subject, as they are, to slow atrophy and resorption."

Influence of Acute Diseases.—Some of the exanthemata, especially variola, scarlatina, measles, and erysipelas, are supposed by Lewin to have a special influence in the pro-

duction of laryngeal polypi; but it is probable that they
only act indirectly by giving rise to chronic inflammation of
the lining membrane of the larynx.

It seems likely also that both croup and whooping-cough
act in the same manner. As regards croup, however, it must
be observed, that though attacks of that disease are likely
to excite growths in the larynx, yet the growths themselves
give rise, in children, to symptoms closely simulating croup.
Hence a previous history of croup must always be looked on
with suspicion. In all cases of laryngeal congestion, but
especially in croup and whooping-cough, it is probable that
not merely the hyperæmia, but the frequent and violent
coughing which characterizes these conditions, has an important causative influence.

*Influence of the Inspiration of Irritating Vapours, and
Particles of Matter.*—The influence of an atmosphere impregnated with atomic matter, in the production of disease, has
been recognized since the time of Pliny, and the Roman
bakers,[1] whilst engaged in their occupation, were in the habit
of covering their face with a kind of cloth. At a later period,
Bubbé,[2] Ramazzini,[3] and others, drew attention to this cause
of disease, and in our own time, Holland,[4] Heussinger,[5]
Virchow,[6] Lewin,[7] Headlam Greenhow,[8] and other physicians
have further elucidated the subject. Quite recently, the asso-

---

[1] Pignorius, *De Servis Veterum*, l. ii. Ramazzini, translated by Ackermann,
1780, p. 126.

[2] *Dissert. inaugur., &c.* Halæ, 1721. *Hufeland's Journ.*, vol. xcviii. p. 4.

[3] *Abhandlungen von den Krankheiten der Künstler und Handwerker*, translated by Ackermann, 1780, vol. i. pp. 123, 147; vol. ii. p. 27.

[4] *Diseases of the Lungs from Mechanical Causes, and Inquiries into Conditions
of Artisans exposed to the Inhalation of Dust*, by Dr. G. Calvert Holland.
London, 1843.

[5] *Ueber Anomale Kohlen- und Pigment-bildung.* Eisenach, 1823.

[6] *Anatomische Beschreibung der Krankheiten der Circulations- und Respirations-
organe.* Leipzig, 1841.

[7] *Beiträge zur Inhalations-therapie in Krankheiten der Respirations-organe.*
Berlin, 1863. The whole subject has been worked out with great industry and
ability, as well as with much statistical evidence, by this author. For several
of my references I am indebted to his interesting work.

[8] *Chronic Bronchitis.* London, 1870.

ciation of "Dust and Disease" has been popularized by Professor Tyndall.[1] The observations of these various physicians and physicists have principally had reference to the production of disease in the lungs and bronchial tubes; but if particles of matter can, as has been so clearly shown, give rise to disease in the lungs, *a fortiori* will they be likely to produce it in the larynx. In the causation of growths, the influence is probably indirect; hyperæmia of the mucous membrane being the immediate result of the inhalation of irritants. Out of 55 adult males in my 100 cases, 8 were engaged in occupations which, of necessity, exposed them to these irritating influences. The employments in the cases referred to were those of carpenter, twine-spinner, wool-packer (2), mason, engine-driver (2), and farrier. In the 105 cases in Appendix D, in which the occupation of adult males is stated, 8 cases occurred in which the patients were subjected to the influence of dusty particles, the occupations being those of lapidary, cabinet-maker, carpenter, mason, tinsmith, chimney-sweep, ragman, fireman, and weaver. It will be seen, therefore, that in 11·25 per cent. of all cases, the patients have been more or less exposed to the mechanical influence now under consideration. It is very probable, also, had the occupations been stated in all the foreign cases, that the influence of molecular irritants would have been still more marked. It must not be forgotten also, that people of every occupation, and of both sexes, are constantly breathing an air impregnated with atomic matter.

Influence of Age.—Dr. Tobold[e] remarks that the affection is most common in middle life, from the 30th to the 60th year, and that laryngeal polypi are least frequently seen in childhood. Dr. Causit,[3] on the other hand, considers that they most frequently occur in early infancy. The latter author, indeed, believes that the disease is very often congenital.

[1] *Dust and Disease;* a Lecture delivered at the Royal Institution, London, 1870.
[2] *Die chronischen Kehlkopfs-krankheiten.* Berlin, 1866, p. 200.
[3] *Études sur les Polypes du Larynx.* Paris, 1867.

The congenital origin of these growths, though very probable, cannot, however, be said to have been established, because in the fatal cases which have been brought forward in support of this view, the patients did not die until they were a year or two old ; and where a laryngoscopic examination has verified the existence of a growth, the little patient has always attained the age of three years or more.

It is very probable, that cases of congenital neoplasm in the larynx do occur, but there is not a single case on record where a still-born child has been found to have a laryngeal growth, nor has such a growth been found to exist within the first month or two of infant life. When a mother or nurse is questioned as to whether a child "sounded its voice" or "cried" in its earliest days, an answer is likely to be given which is not only evidently influenced by the present condition, but which also shows an entire ignorance as to the difference between articulation and vocalization. If a child has never spoken, that is, never articulated vocal sounds, the ignorant are apt to say "the child never had any voice," even though it may have been in the habit of screaming lustily. There are cases, indeed, where the mother can be made to understand the essential difference between articulation and vocalization, as in cases reported by myself;[1] but even in such cases, though congenital aphonia is proved, the positive cause of that aphonia cannot be ascertained, and the condition of the larynx at an early period may have been merely that of hyperæmia.[2]

According to my experience, the middle period of life would appear most favourable to the development of these neoplasms, and I find that after the age of 50 there is a considerable and sudden diminution in their number.

[1] Appendix A, Case 98 ; and Appendix B, Cases 3 and 9.

[2] In one very remarkable case reported by Dr. Causit (*Op. cit.*, p. 82), an infant, at the moment of its birth, only made a very weak mewing sound (*miaulement très faible*). For three weeks afterwards it could only utter faint cries, and never those of a sonorous or piercing character, and after that period the voice ceased completely : the cough was aphonic. The child died when it was about a year old, and a growth was found occupying almost the entire larynx.

In the 100 cases tabulated in Appendix C as having been treated in my own practice, the decennial period in which the greatest number of cases occurs is from 40 to 50, no less than 28 patients having been of that age, whilst there were as many as 72 between the ages of 20 and 50. The following table shows the proportion of cases at different ages :—

Between the ages 2 and 5 there were 2 cases.

5	„	10	„	4
10	„	15	„	4
15	„	20	„	2
20	„	30	„	21
30	„	40	„	22
40	„	50	„	28
50	„	60	„	14
60	„	70	„	3

100

These statistiscal results, based on cases actually treated, do not include two cases of supposed congenital growth contained in Appendix B.

The facts contained in the tabular statement of my own cases are corroborated by an analysis of the ages of the patients in the hands of other practitioners. By reference to Appendix D, it will be seen that of 163 cases in which the ages are stated, no less than 112, that is 68·7 per cent., occurred between the ages of 20 and 50. The greatest age, at which a growth has been seen, occurred in the practice of Dr. Bruns. In one of his cases the patient was 74 years old.[1]

It is probable, however, that the actual number of cases of laryngeal growths in young children is much greater than my tables would indicate ; for it is extremely likely that many cases of growths in the larynx in young subjects are overlooked by practitioners who do not use the laryngoscope ; and, moreover, as in young subjects, the epiglottis is more pendent, growths, though present, may not be revealed even

[1] Appendix D, Case 123

by the laryngoscope. Statistics as to age, mainly founded on the experience of the Hospital for Diseases of the Throat, are subject to a certain fallacy, inasmuch as young children are more likely to be taken to one of the numerous Children's Hospitals, now so conveniently situated in almost every part of London, than to the former Institution. It is worth noting here, that of the 34 morbid specimens collected in the various metropolitan museums, no less than 15 are from children under the age of 12 years.

Influence of Sex.—As to sex, I find that of my 100 patients, 62 were males, and 38 females. Of 187 patients in the practice of other operators, whose sex is tabulated in Appendix D, 135 were males, and 52 females. These numbers show that neoplasms, like other laryngeal diseases, are more common in the male than in the female sex. This is, perhaps, partly due to the fact that, from the nature of their occupation, men are more exposed to the exciting causes of chronic hyperæmia. It does not, however, altogether explain the greater proclivity of the male sex ; for in Causit's 42 cases, in which the growth occurred in young children, before the influence of vocation could modify results, 28 occurred in males, and 14 in females.[1]

Influence of Occupation.—The examination of my tables (Appendix C) would seem to indicate that the professional use of the voice is one of the circumstances most favourable to the development of growths. Thus, if we except the occupations of gentlemen, merchants, and labourers, each of which embraces people subject to very different conditions, though they furnished respectively 2, 2, and 4 cases, it will be seen that a preponderating number belong to those who are constantly obliged to use their voice, no matter what may be the state of the vocal organ. Thus of the 53 males old enough to have an occupation, 6 were vocalists, 1 a clergyman, 2 officers in her Majesty's service, 2 waiters, 1 a page, 1 a sailor, 2 hawkers, and 1 a railway porter whose duty in-

[1] *Op. cit.*

volved constant shouting. There were also 3 females who professionally used their voice ; viz., 2 hawkers and 1 Scripture-reader. These embrace 21 per cent. of all my treated cases old enough to have an occupation.[1] Of 105 cases (of the 189 tabulated in Appendix D) in which the occupation is stated, no less than 20 were people whose profession entailed the constant use of the voice. As a proof, however, that growths may arise from causes entirely independent of vocalization, attention is called to Case 37 in Appendix D, in which the patient was a deaf mute.[2] The influence of occupation, in so far as it subjects people to the inspiration of irritating particles, has been already considered (pp. 12 and 13).

Of my treated cases, 23 were engaged in out-door occu-pations,[3] 70 in in-door occupations, and in 7 the employment was of a mixed character. When it is remembered that the actual number of people employed out of doors in large towns is very much less than those engaged in in-door occupations, and also that, according to the general statistics of the Hospital for Diseases of the Throat (in which the majority of my cases occurred), those engaged in-doors furnish by far the greater number of cases,—it is all the more remarkable that laryngeal growths should be more prevalent among those employed in out-door occupations. This difference in the statistical conclusions may, perhaps, be explained by the fact that growths most frequently arise where *people use the voice out of doors*, as in the case of mili-tary and naval officers, hawkers, street-singers, &c.

[1] As a matter of convenience, for statistical purposes, patients have not been considered to have an occupation before the age of fifteen years.

[2] In referring to this case, I do not mean to imply that deaf mutes are entirely incapable of any vocalization, but that in their case vocalization is quite a secondary and unimportant function.

[3] Of course, most of the women and children, and some of the men, had " no occupation," and in considering the respective influence of external and internal atmospheric conditions, deductions are based on the supposition of the patient spending the greater part of his or her time either in-doors or out-of-doors.

3

SECTION III.

—◦◦◦—

SYMPTOMS.

It will be readily understood, that, as a rule, the signs and symptoms of a growth in the larynx depend in their nature and degree upon the exact situation and size of the neoplasm. Thus a growth on the vocal cords causes aphonia, or hoarseness ; a growth on the epiglottis produces dysphagia ; and a large tumour, wherever situated, is likely to give rise to dyspnœa.

Symptoms are *functional*[1] [alterations of voice, dyspnœa (including stridor and paroxysmal suffocation), pain, and difficulty of swallowing], and *physical* (furnished by laryngoscopic examination, by the laryngeal sound, by digital exploration, by forcible external elevation of the larynx combined with depression of the tongue, by auscultation and percussion of the larynx, by examination of the expectoration, and by the constitutional conditions).

Functional Signs.

Functional signs furnish very imperfect evidence, except to those who have had large experience of the cases under consideration. From the varying and peculiar character of the voice, the croupy cough, and the paroxysmal dyspnœa, the presence of a growth may be occasionally inferred, by the experienced laryngologist ; but those who have not met with many laryngeal polypi would be rash to form a diagnosis from functional symptoms. It must not be forgotten, however,

---

[1] This arrangement of the symptomatology, I have modified from Dr. Causit's most valuable work, already referred to. It appears to me to be preferable to the divisions of symptoms into *subjective* and *objective*, which I formerly employed.

that many years before the laryngoscope was invented, both Brauers and Ehrmann [1] were able to diagnose growths with such accuracy, that they felt justified in opening the thyroid cartilage. In Brauers' case, the use of the actual cautery brought on hectic fever, but Ehrmann's skilful diagnosis and bold treatment, as is well known, were crowned with success.

Modification of the Voice.—An alteration in the voice is the most constantly present, though not invariable, symptom of a growth in the larynx. Out of my 100 tabulated cases, the voice was impaired 90 times; there being complete loss of voice in 55 cases, and hoarseness in 37 cases. Of the 55 cases in which there was complete aphonia, this was the only symptom 35 times, other symptoms being also present in the remaining 20 cases; whilst of the 37 cases in which dysphonia occurred, the alteration of voice was the only symptom 17 times. It will be seen, therefore, that impairment of the voice was the unique symptom in no less than 52 per cent. of my cases. Aphonia or modification of the voice was present in 92 of the 171 cases in Table D in which the symptoms were stated, that is, in 53·8 per cent. of the cases.

There is a kind of dysphonia which, when present, is very characteristic of growth-cases. The patient whilst speaking in his natural voice, or in a slightly hoarse or croupy tone, suddenly becomes completely aphonic, and again, after a minute or two, recovers his hoarse or natural voice.[2] The only kind of dysphonia which at all resembles this, is found in a form of nervo-muscular disease of the larynx, in which the tensors of the vocal cords are spasmodically affected. The constant straining character of the voice in these latter cases, however, at once differentiates them from all others.

In the history of these cases we generally find that dysphonia precedes complete loss of voice; but, as the growths

---

[1] *Op. cit., Observations,* Cases xv. xxix.

[2] Krishaber (*Op. cit.*) has applied the term *vocal asynergy* to the loss of power of control over the voice, which is so often present, and indeed one of the first symptoms of a growth in the larynx. As, however, *asynergy* is a term implying a want of correlation, dependent on impaired nervo-muscular power, it is evidently incorrect to apply the word, in cases where the absence of correlation is due to direct mechanical causes.

originate in hyperæmia, the early hoarseness may, in some cases, be due to that condition. Sometimes the growth does not impair the ordinary speaking voice, but entirely destroys many of the notes in singing,[1] and in the case of vocalists the hoarseness is sometimes also preceded by a want of power and of control in the higher notes.[2] In some cases the natural voice of the patient is thick, but he is able to speak in a clear falsetto voice. A double-toned vocalization, in which (secondary) falsetto sounds are heard simultaneously with the ordinary, though impure, voice, has been described by Türck as Diphthonia.[3]

The *modus operandi* of these growths is manifold : sometimes by their situation on the vocal cords, they directly modify the vibration of the cords ; sometimes, by their pressure on the ventricular bands, they interfere with the

---

[1] Appendix A, Cases 26, 39, and 77.

[2] *Ibid.*, Case 93.

[3] Dr. Türck relates, amongst others, the case (Case 206, *Op. cit.*) of a man, aged 44, who had small growths on both vocal cords :—" The chest voice was very hoarse, and his intonation very impure. His ordinary speaking voice reached from *f* to *c*, but he could go down as low as *c* below the line. In loud vocalization and loud speaking, the first *a* above the line sounded simultaneously as a falsetto. In the lower (chest) notes, the associated falsetto note (*a* above the line) was more often present than in the higher notes. In all the various notes which he could produce, the high falsetto note remained entirely unaltered ; in phonetic inspiration there was no double tone." Merkel (*Anatomy and Physiology of the Human Vocal Organs*, Leipzig, 1857, p. 628, quoted by Türck, *Op. cit.*, p. 473) observes that " when the voice is, so to speak, slightly veiled, a little mucus having remained in the larynx, between the vocal cords and ventricular bands, if one tries to pass from a clear high falsetto gradually down, it often happens that these tones sound impurely. In my own case, it occurs most frequently, when *b* has been the fullest and loudest falsetto tone (sometimes it is *b* flat or *a*), that the next three tones, *a*, *g*, and *f*, show this impurity ; that is to say, they are accompanied by an intervening sound or jarring incidental tone, which, as a rule, is exactly an octave lower than the principal tone. On one occasion, the incidental tone was almost as loud and distinct as the principal tone, so that I could sing a whole passage in the right octaves, although they did not sound quite harmonious. This phenomenon was seldom shown in more than three consecutive tones. It is true that it sometimes happens that the impure incidental tone belongs to another scale, but this is more rare." The vocal phenomena, here described by Merkel, as occasionally resulting from the accidental lodgment of mucus in the neighbourhood of the vocal cords, occur as a more permanent condition when produced by a growth.

motion of air in the ventricular cavities; sometimes, when broadly attached, and projecting into the centre of the laryngeal canal, they prevent the vocal cords being thrown into vibration by the expired current of air; and sometimes they directly interfere with the approximation of the cords. Growths situated near the centre (of the antero-posterior diameter) of the vocal cords alter the voice, *cæteris paribus*, to a greater extent than those attached at the extremities; but those actually springing from the anterior commissure, by preventing the approximation of the cords, obliterate the voice even more promptly and completely. As has been remarked by Czermak, a small growth often interferes with vocalization more than a large one; for the small neoplasm, which is almost always sessile, greatly modifies the vibration of the vocal cord to which it is attached, whilst a large one may become pedunculated as it grows, and by rising up into the cavity of the larynx, may interfere very little with the normal formation of sound. Growths on the epiglottis and ary-epiglottic folds do not generally affect the voice, unless they attain a very large size; and small neoplasms on the ventricular bands do not always cause dysphonia. Growths below the vocal cords, by diminishing the column of air passing through the larynx, or by being forced up into the glottis in expiration, often cause aphonia.

Cough.—Patients with laryngeal growths do not, as a rule, suffer much from cough; but, occasionally, on the other hand, this symptom is so severe as to cause very great inconvenience, and it may even give rise to hæmoptysis. In Dr. Causit's 46 cases of laryngeal growths, 26 of the patients suffered from cough; but, in my 100 cases this symptom was troublesome in only 12 instances. In Appendix D there are 171 cases in which the symptoms are stated, and in 27 cough was present, that is, in 15·78 per cent. The character of the cough depends upon the size and situation of the growth; it is generally dry and hacking, and often aphonic. In young children or in adults, when the growth is very large and situated in the neighbourhood of the glottis, it has often a croupy character. In 7 out of

the 26 cases noticed by Dr. Causit, it was described as
"croupal." I have seen it occur also in two cases in violent
paroxysms.

Dyspnœa.—Dyspnœa was present 30 times in my 100
cases, and was serious or threatened suffocation in 15 cases.
Difficulty of breathing occurred in about the same percentage of the cases I have collected in Appendix D. Most
of the specimens of laryngeal growths in the metropolitan
museums were taken from patients who died from suffocation ; and in nearly all the cases reported in the medical
journals before the invention of the laryngoscope, dyspnœa was a prominent symptom. The dyspnœa is often
paroxysmal. The explanation of this circumstance, as in
many other cases of laryngeal obstruction, is, that the
patient is able to breathe well, even through a narrowed
windpipe, provided that no further diminution suddenly
occurs. If, however, the patient takes cold, and the mucous
membrane becomes a little swollen, a paroxysm of dyspnœa
may supervene. In the same manner, if the respiration
be hurried by exertion, an attack is likely to come on.
Dyspnœa likewise sometimes occurs suddenly, from the
patient getting into an unusual position, and from the growth
being consequently thrown more across the glottis. In one
of my cases[1] the patient *could only sleep with the hand
resting under the neck ;* and if by chance her head slipped
away during sleep, she immediately woke with a severe
attack of dyspnœa.

As has been pointed out by Dr. Causit, in analyzing his
collected cases, the attacks of suffocation most frequently
occur in the evening and during the night. This circumstance, however, rather increases than diminishes the difficulty of estimating the value of the symptom in diagnosis,
as in almost all affections attended with dyspnœa, exacerba-
tions take place at night. The *stridulous* character of the
dyspnœa differentiates it, however, from the various forms of
asthma. The result of my experience enables me to entirely
endorse the correctness of Dr. Causit's observations, as to

[1] Appendix A, Case 84.

the mode in which the paroxyms occur :—" The attacks of suffocation are, in general, rare at the commencement, but as the disease advances they increase in intensity and frequency. The intervals are at first very long (from some weeks to months, or years), but they gradually shorten, so that the attacks become very numerous and very intense." As a rule, the amount of dyspnœa depends on the situation of the growth and on the relation of its size to that of the laryngeal canal ; but this is not invariably the case : in one case[1] that has come under my notice, a subglottic growth of moderate dimensions caused extreme dyspnœa by preventing the abductive movement of the vocal cords, and some of the largest growths which I have met with have caused but little embarrassment in the breathing ; whilst small excrescences sometimes give rise to violent spasm or persistent dyspnœa.[2] It almost invariably happens, that inspiration is much more difficult than expiration, and Lewin[3] has remarked, that the character of the respiration has a certain diagnostic value, as regards the seat of the growth. When inspiration is noisy and stridulous, and expiration comparatively easy, the growth is probably situated above the vocal cords, and *vice versâ.*

Pain.—According to my own experience, actual pain is seldom caused by growths in or about the larynx, but uneasy sensations are occasionally felt. In only one of my 100 cases[4] was there actual pain, and in one[5] there was a sensation of oppression. In the foreign tables,[6] however, pain[7] is stated to have occurred in no less than 9 cases ; 7 patients experienced the sensation of a foreign body in the larynx, and in 3 cases there was a feeling of oppression. Though, however, patients rarely complain of the feeling of a foreign body, they frequently have a disposition

---

[1] Appendix A, Case 81. [2] Appendix A, Cases 21 and 60.
[3] *Deutsche Klinik*, 1862. [4] Appendix A, Case 97.
[5] Appendix A, Case 90. [6] Appendix D.
[7] Although I have endeavoured to exclude all cases of cancer from Appendix D, it is possible that some malignant cases have been accidentally inserted. If this has occurred, it will account for the greater frequency of pain in the cases of other practitioners than in my own cases.

to clear the throat, as if to expel some accumulated mucus. I have most commonly met with this symptom in cases of pedunculated growths, especially when they were attached to the vocal cords. In one case,[1] in which there was a growth the size of a pea on the cartilaginous portion of the right vocal cord, the only symptom complained of was a sensation " of a constant tickling in the throat ; " and the same sensation was experienced in another case, in which the growth was situated in the hyoid fossa.[2] In one case, in Appendix D, there was simply a tickling sensation in the larynx, and in another, great irritability of the palate.

Dysphagia.—Difficulty of swallowing does not generally occur, except where the growth springs from the epiglottis or where it attains a very large size ; it is occasionally present, however, when the neoplasm arises from the arytenoid cartilages. In my 100 cases here tabulated (Appendix C), dysphagia was only present 8 times, and in each of them, with perhaps one exception,[3] the epiglottis was the seat of the disease. In the exceptional case referred to, there was difficulty in ascertaining the exact origin of the growth: it was thought to spring from the ventricular band ; but in all probability the epiglottis was also involved. In one case only[4] was there odynphagia,[5] i. e. *pain in swallowing.*

PHYSICAL SIGNS.

The physical signs are much more important than those of a functional character ; and amongst them those observed with the laryngeal mirror stand pre-eminent.

---

[1] Appendix A, Case 13.   [2] Appendix A, Case 89.

[3] Appendix A, Case 83.   [4] Appendix A, Case 28.

[5] Impaired deglutition may be due to *difficulty* of swallowing, or *pain* in swallowing : both these symptoms are commonly known under the name of dysphagia (ὅυς, *difficulty*), which term should be limited to those cases in which there is obstruction or loss of power. Where, however, impaired deglutition is the result of pain, odynphagia (ὀδύνη, *pain*) would be more correct. This is not a pedantic technicality, as deductions, drawn from the use of one or other of these terms, would entirely depend on which term were employed.

Laryngoscopic Signs.[1]—So complete is the information furnished by the laryngoscope, that were it not that there are certain rare and exceptional cases in which this instrument cannot be employed, the general semeiology would be useless. The situation of the growth can almost always be ascertained with the mirror, but in a few cases, where the growth is very large, the *exact seat of origin* may be concealed. The following table shows the parts most frequently affected in my 100 (treated) cases :—

Both vocal cords	27	times.
Right vocal cord	25	,,
Left vocal cord	14	,,
Vocal cords and ventricular bands	2	,,
Vocal cords and epiglottis	4	,,
Epiglottis .	8	,,
Epiglottis and ventricular band	1	,,
One or both ventricular bands .	6	,,
Posterior wall of larynx .	7	,,
Whole surface of larynx .	3	,,
Capitulum Santorini .	1	,,
Inter-arytenoid fold .	1	,,
Hyoid fossa	1	,,

It is thus seen that the vocal cords are especially liable to be affected, these parts having been alone attacked in 74 cases, and suffering in conjunction with other parts in no less than 85 cases. Of the cases in Appendix D, in which the situation is stated, the vocal cords were alone affected in 61·4 per cent., and in conjunction with other parts of the larynx, in 64·4 per cent. On the other hand, it will be seen the arytenoid cartilages, with their folds of mucous membrane and secondary cartilages, enjoy great immunity.

The laryngoscopic appearance can best be described in detail, by separating the different kinds of tumours, according to their pathological nature.

Papillomata.—Papillary growths are generally sessile,

[1] Under this head, the form, colour, size, and situation of the various growths will be considered ; the information as regards structure will be found in the section on Pathology.

4

though occasionally pedunculated. They are often multiple, and sometimes occur symmetrically.[1] They vary in size, from a grain of mustard to a walnut, but they do not often attain the latter dimension. Their most common size is that of a large split pea. These growths may have a mammillary, cauliflower, raspberry, foliated, fimbriated, dentated, or vermiform configuration. They are generally of a pink colour, but they may be white, or even bright red, as will be seen by reference to Plate II. figs. 1, 2, 3, 5, 7, and 11.

In no less than 42 of my 67 cases of Papilloma, the disease was confined to the cords, and in 6 instances the vocal cords were affected in common with other parts of the larynx. It will thus be seen that the vocal cords were implicated in more than 70 per cent. of the cases of Papilloma. Of the 42 cases in which the vocal cords were alone affected, the right cord was 16 times the site, the left 7 times, and both cords 19 times.

Benign Epithelial Growths.—Benign epithelial growths (Plate V. fig. 1) are generally sessile, and vary in size from a split tare to a sparrow's egg. Their surface is generally smooth, but they may be furrowed or even lobulated. They are commonly white or pale red. They are most frequently seen upon, or very near, the vocal cords.

Fibromata.—True fibromata (Plate II. fig. 4, and Plate III. fig. 10) are usually round or oval, but occasionally are of a very divided form, not unlike cauliflower excrescences,[2] and are generally, but not invariably, pedunculated. Their surface is usually smooth, but it may be rough, irregular or wavy. They are commonly of rather a bright red colour. They are almost always single, and vary in size from a split pea to an acorn. Rokitansky,[3] however, has reported one case, in which the growth was as large as a pigeon's egg. These growths, like the papillary tumours, occur most frequently on the vocal cords. In 5 of the 11 cases now given, the cords were the sole seat of the neoplasm.

Fibro-cellular Growths.—Fibro-cellular tumours (Plate II.

---

[1] Appendix A, Cases 40 and 80.

[2] Appendix A, Cases 78 and 97.   [3] *Op. cit.*

fig. 10) are almost invariably pedunculated, and of a round, or pyriform, contour. The colour may be either pink, or bright red, and the surface is generally smooth. Unlike mucous polypi of the nose, these growths are generally single. In one case [1] the growth was as large as a cherry, but the others were very small pyriform tumours. In 2 cases, the growths were attached to the epiglottis, and in the 3 others to the vocal cords.

Myxomata.—These growths are very rare. In the single case [2] (Plate III. fig. 11) which I have met with, the neoplasm grew from the right vocal cord, and was only in part of a mucous character; this portion was seen with the laryngoscope to be quite transparent and of a bright pink colour. In a case reported by Dr. Bruns, the growth which, like mine, was on the right vocal cord, was about the size of a filbert slightly lobulated, of pale pink colour, and smooth surface; before removal it was regarded as a "soft fibroma."

Lipomata.—No case of lipoma has come under my own observation, but Professor Bruns [3] has met with one instance. The growth, which was very large, obscurely lobulated, and of a bright red colour, sprang from the mucous membrane covering the left arytenoid cartilage, and occupied almost the entire larynx from before backwards, as well as from side to side. The examination of the growth enabled Dr. Bruns to *diagnose its nature* whilst it was *still in situ.* It was ascertained to be soft and elastic; the laryngeal sound could be easily *pressed into* the tumour, and on withdrawing it, the impress of the sound immediately disappeared.

Fasciculated Sarcomata (Plate III. figs. 3, 7, and 8).—There is nothing distinctive in the appearance of these growths. They are sometimes rough and sometimes smooth, sometimes white and sometimes pink or red. In one of my cases the growth sprang from the vocal cord, in another, from the anterior commissure, and in a third, from the ventricular band. In 4 of the 6 foreign cases, the growth was situated in the neighbourhood of the vocal cords, and in the remaining two the disease was general.

[1] Appendix A, Case 52.   [2] Appendix A, Case 99.
[3] Appendix D, Case 156.

Cystic Growths.—Cystic tumours (Plate II. fig. 6, and
Plate III. figs. 1 and 2) are round, egg-like projections, and
as they usually give rise to some local irritation, they are
themselves red, and are surrounded by a hyperæmic area.
In one of the cases [1] hereafter reported, the growth was the
size of a sparrow's egg; in the other [2] it was as large as a
cherry. In both my cases, as well as in one recorded by
Mr. Durham,[3] the tumours grew from the epiglottis; but
cystic growths, springing from the ventricle, have been reported by Gibb,[4] Bruns,[5] and others.

Adenomata.—There is nothing characteristic in the appearance of glandular tumours in the larynx. In one of my cases,
situated below the anterior commissure of the vocal cords,
the growth looked like an ordinary cauliflower excrescence,
and in the other the tumour, which grew from the epiglottis,
was large and nodulated, and had very much the appearance
of an hypertrophied tonsil (Plate III. figs. 4, 5, and 6). In
both cases, the growth was of pink colour.

Angeiomata.—The only case of vascular growth [5] (Plate II.
fig. 12) which has come under my notice, was of a black-
berry-like appearance, in colour, form, and, size, and grew
in the right hyoid fossa.

Laryngeal Sounds and Crotchets.—By means of the laryngeal sound or probe, the density, the size, and the exact origin
of a growth may often be determined, when with the laryngeal
mirror alone there is still doubt as to these various points.
A smooth growth may be either a fibroma or a lipoma; but
whilst the former does not yield to pressure, the fatty growth
is soft and resilient. The appearance of a laryngeal growth
in the mirror is often deceptive, and it is often only by
moving it with the sound, that its dimensions can be at all
accurately determined. This is more especially the case,
because, of course, only one surface of the tumour is visible
in the mirror. Again, the insertion of a growth is sometimes

---

[1] Appendix A, Case 25.   [2] Appendix A, Case 85.
[3] *Medico-Chirurgical Transactions*, 1863; and Appendix D, Case 55.
[4] Appendix D, Case 57.   [5] Appendix D, Case 84.
[6] Appendix A, Case 89.

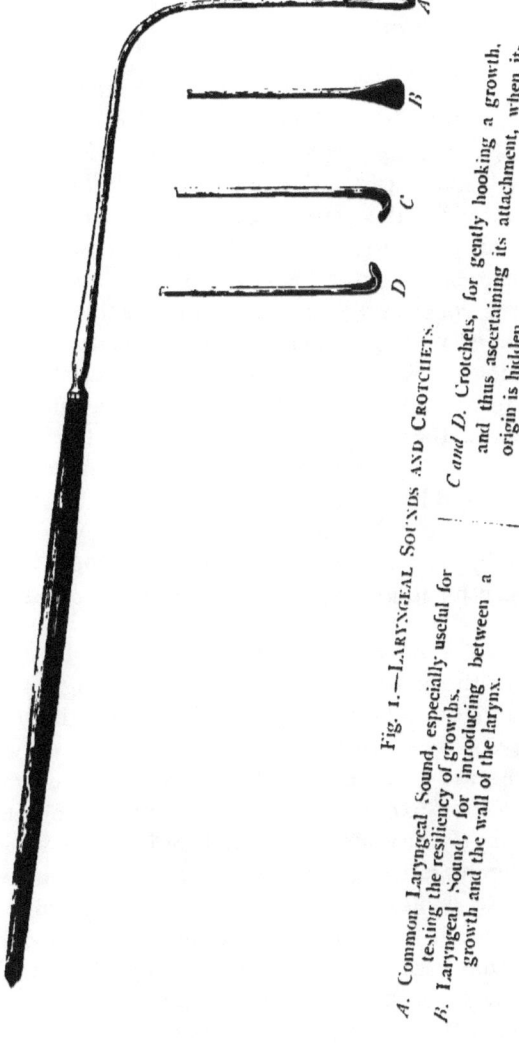

Fig. I.—LARYNGEAL SOUNDS AND CROTCHETS.

A. Common Laryngeal Sound, especially useful for testing the resiliency of growths.

B. Laryngeal Sound, for introducing between a growth and the wall of the larynx.

C and D. Crotchets, for gently hooking a growth, and thus ascertaining its attachment, when its origin is hidden.

hidden by the growth itself, and it is only by traction with
the crotchet that the precise origin can be ascertained. The
various kinds of sounds and crotchets which I am in the habit
of employing are shown in Fig. 1.

Digital Exploration.— Examination with the index-finger
is of some value, in those cases, where the growth is situated
on the epiglottis, or the ary-epiglottic folds; but it is seldom
of any practical service, where the tumour is attached at a
lower level. By means of this method, evidence can sometimes
be obtained, as to the density and mode of insertion of the
growth. In young children, the finger can, indeed, be passed
down as far as the vocal cords; but, as in these subjects, the
growths are generally of extremely soft, papillomatous character, they cannot, unless large, be detected with the finger.
It is important to bear in mind, that in these young patients,
before the cornua of the hyoid bone are developed, the body
of the bone feels, on introduction of the finger, very much
like a hard growth, and I have reported one such case[1] in which
a mistake actually occurred. In this instance, a growth was
present; but on digital examination, the body of the hyoid
bone was mistaken for it, and the growth itself was not felt.

Forced External Elevation of the Larynx.—By pressing the
larynx upwards with the hand on the thyroid cartilage, and
by pulling the tongue out, the upper orifice of the larynx
may occasionally be seen. In this way, growths in the upper
part of the larynx are sometimes visible. Voltolini[2] recom-
mends that in addition to external manipulation, and holding
out the tongue, the fauces should be slightly irritated, so as
to produce moderate retching. By adopting this procedure,
he was enabled to demonstrate a growth on the posterior
wall of the pharynx just at the orifice of the œsophagus, to
Professor Middeldorpf, who succeeded in removing the tumour
with a galvanic cautery loop, and (though an accurate diagnosis was first made with the laryngoscope) in carrying out the

[1] Appendix B, Case 3.
[2] *Separat-Abdruck aus der Berlin. Klinik. Wochenschr.*, 1868, No. 23.

treatment the laryngeal mirror was not used. Other cases [1] have occurred, where large growths in this situation, have been seen projecting into the pharynx, when the mouth was widely opened, without any external manipulation of the larynx.

Auscultation and Percussion of the Larynx.—On auscultation of the larynx when the growths are at all large, moist sibilant râles may be sometimes heard, but they are only characteristic of laryngeal obstruction. On this subject, Rühle [2] remarks, "that in pedunculated tumours of the larynx an auscultatory phenomenon, a valvular murmur is sometimes heard, which, by the exclusion of the accidental impaction of a foreign body, and by its repeated occurrence during a considerable length of time, has undoubtedly a certain value." Undue importance has been attached to this valvular sound, which took its origin in Ehrmann's celebrated case, [3] in which the patient, by a sudden inspiration and expiration, "was able to imitate the sound of a valve alternately opening and shutting." This phenomenon, which was quite accidental, could be produced by a voluntary effort on the part of the patient.

When the larynx is blocked up with growths, dull sounds are elicited on percussion. Small growths, however, do not in any way modify the usual resonance.

Microscopical Examination.—It occasionally happens, especially in papillomatous growths, that small particles are expectorated, and, on microscopical examination, their nature can be verified. When this occurs in conjunction with other symptoms, it of course furnishes general evidence as to the nature of the disease ; and when there is aphonia at the same time, it may be inferred that the growth is in the neighbourhood of the vocal cords. This evidence of the presence of growths is, however, so rarely present, that it must be

---

[1] *Polypi of the Larynx*, by Dr. Horace Green. New York, 1852, p. 62. Also Rayer, *Maladies de la Peau*, tom. ii. p. 422.

[2] *Op. cit.*, p. 223.

[3] *Op. cit., Observations*, xxix. p. 23.

regarded more as an accidental phenomenon, than a sign of
the disease. These remarks on the microscopic investigation of growths only apply to the examination of particles
accidentally expectorated. The histological differences of
the various laryngeal growths are described in detail, in the
section on Pathology.

Constitutional Condition.—In the early stages, the disease
is purely local; but if the growth become large, it may, by
embarrassing the respiration, or through other causes, give
rise to constitutional disturbances; in this way, some amount
of wasting and hectic may be caused, so that these cases
were formerly mistaken for phthisis. Marked constitutional
symptoms are, however, of exceedingly rare occurrence.

Course and Termination.—The various symptoms already
described generally develop themselves slowly, taking many
months for their evolution. There is always a difficulty,
however, in fixing upon the commencement of the disease,
because the hyperæmia, which generally precedes the growth
of a tumour, gives rise to the same phenomena as the neoplasm itself. The progress of the case depends, of course,
in a great measure, on the pathological nature of the neoplasm. In four instances,[1] however, out of my 100 treated
cases, I think the date of origin may be at least limited to
within a certain period. In one case, a growth which blocked
up the entire larynx was certainly of not more than six
months' duration; in the second case, two months previous
to the discovery of a growth, the size of a pea, on each vocal
cord, I had examined the larynx, and had noticed only congestion of the mucous membrane of the cords; in the third case
a very large growth was seen after an interval of nine months
from the date at which the patient had left my care, just
recovered from a sharp attack of laryngitis; in the fourth case
the patient left me with only slight thickening and a tendency
to ulceration of the epiglottis, on November 6th, 1869, and
within two months, returned with a growth the size of a large
cherry.

[1] Appendix A, Cases 63, 80, 86, and 88.

After attaining a moderate degree of intensity, the symptoms often remain stationary, and it is surprising how long patients—especially those among the lower classes—suffer from aphonia before they seek relief. In one of my cases the patient had suffered from aphonia for 24 years, and another from dysphonia for 23 years. Dyspnœa, however, is a much more serious and a more progressive symptom, and after a few months, the patient is obliged to apply for medical aid. Dysphagia also, being a constant source of annoyance, quickly leads the sufferer to seek relief. If the growth continue to increase, and be not checked by treatment, the case is likely to end in suffocation. The termination of the disease is more fully considered under the head of Prognosis.

As a curious fact recorded in medical literature, rather than a practical matter bearing on the course of laryngeal growths, it may be remarked, that there are a few instances in which the disease has been cured spontaneously. In one of these, briefly referred to by Causit,[1] the polypus was expelled by coughing. In another case, reported by Dr. Dobie,[2] a pedunculated growth, about the size of a small cherry, attached to the epiglottis, separated spontaneously. Türck[3] also relates a case, in which this fortunate termination took place as the result of acute laryngitis.

Complications.—Complications are fortunately rare in the natural history of laryngeal growths. The disease being undoubtedly of local character, complications which arise, are principally due to mechanical causes. Spasm of the glottis is essentially a direct *consequence* of a growth, but œdema of the glottis, which is only an occasional concurrent, may fairly be regarded as a *complication*. Krishaber[4] remarks that in one autopsy he found considerable pulmonary emphysema, and cardiac hypertrophy.

[1] *Op. cit.,* p. 33.
[2] *American Monthly Journal of Medical Science,* 1853.
[3] Türck, *Klinik der Krankheiten des Kehlkopfes,* p. 305. Wien, 1866, p. 305.
[4] *Op. cit.,* p. 749.

SECTION IV.

—◆—

DIAGNOSIS.

GROWTHS in the larynx cannot well be mistaken for any other disease, if a laryngoscopic examination be possible.

Eversion of the Ventricle is, perhaps, the only intelligible source of error, and this condition is probably rare. I know of only two specimens; one of these was exhibited by Dr. Moxon, at the Pathological Society;[1] the other is in the museum of the Hospital for Diseases of the Throat. Dr. Moxon's specimen was removed from the body of a man who died in Guy's Hospital from cancer of the stomach. Dr. Moxon had often spoken to the patient, whose " *voice was always such as not to attract attention.*" After death, a tumour was found hanging down over one of the vocal cords. It was semi-elliptical in shape, and was rooted above in the anterior half of the ventricle of the larynx. The tumour " could be easily put up into the usual position of the sacculus laryngis ; that when so placed—the tumour inverted, and returned behind the false vocal cord—it appeared as the sacculus laryngis, while, without it, there was no sacculus laryngis ; so that there could be no doubt that it was an everted sacculus." My specimen was taken from a patient who had been admitted into the Hospital on the night previous to his death, in a very debilitated condition; no notes were taken, and there is therefore no record as to the state of the voice. After death, the left ventricle of the larynx was found to be entirely everted, and the right sacculus protruded slightly from the

[1] *Transactions of the Pathological Society*, vol. xix. p. 65. The specimen is also described and figured by Mr. Durham (*Op. cit.*, p. 580).

ventricular orifice. The appearance is shown in Plate V.
fig. 2. On carefully sawing out a portion of the left ala of
the thyroid cartilage from without, it was seen that there
was no sacculus laryngis on the left side; but on return-
ing the protruded sac, the well-known appearance of the
Phrygian cap was presented to view, and the anatomical
aspect of the larynx on this side became perfectly normal.
The mucous membrane of the larynx was seen to be covered
with superficial ulcers: the destructive process appeared to
have principally affected the glandulæ, and was most marked
on the ventricular bands, and the cushion of the epiglottis.
There were cavities in both lungs.

In Dr. Moxon's case, there is every reason to believe that
the prolapse took place *in articulo mortis*, and it is quite
possible that such may have been the case in my own
specimen. Dr. Moxon remarks that this condition would
"be very tempting to one skilled in the removal of laryngeal
polypi," and indeed, if eversion took place during life, I know
of no other treatment but excision of the sac which would
relieve the symptoms.

The diseased conditions which might be mistaken for
growths, are those occurring in syphilis and laryngeal phthisis;
but independently of their commemorative signs, these
diseases are easily differentiated by means of the laryngeal
mirror. Elephantiasis is so rare in this country, that it only
requires a passing word, and in lupus, loss rather than
increase of substance is the distinguishing feature. It is, however, more important to point out the differences of malignant
tumours and outgrowths.

Syphilis.—The *condylomata* of syphilis are seen as irregular,
whitish, very slightly raised, prominences on the congested
membrane, the posterior wall of the larynx being their most
common site. These formations are comparatively rare, and
when present, generally occur from six weeks to three months
after the primary inoculation. ·They soon disappear under
the use of mineral astringents, or other remedies which modify
the nutrition of the part, and, even if left to their natural
course, they quickly subside.

False Excrescences are the result of syphilitic ulceration and subsequent cicatrization, and occur as irregular projections in different parts of the larynx. Though seldom, in themselves, causing dyspnœa, this symptom may result from the narrowing and distortion of the windpipe, which takes place from the accompanying cicatricial contractions. False excrescences are often accompanied by dysphonia or complete loss of voice ; but though their removal sometimes gives considerable relief, it does not generally restore the vocal function, owing to the associated pathological conditions already referred to.

The large roundish *gummata*, which are occasionally found in the larynx, are so closely incorporated with the adjacent tissues, that they are not likely to be mistaken for true laryngeal growths.

Elephantiasis.—In the few cases of elephantiasis that have come under my notice, in which the larynx was affected, the mucous membrane covering the epiglottis was uniformly swollen.[1] I believe that the disease never attacks the mucous membrane until after it has shown itself on the tegumentary surface.

Lupus.—The thickening of lupus is generally very much like that which occurs in tertiary syphilis, and is usually soon followed by destructive ulceration.

Laryngeal Phthisis.—The thickening of laryngeal phthisis has not the defined character of a true laryngeal growth, and is generally soon followed by ulceration.

Malignant Growths.—It is not always easy to distinguish between benign and malignant laryngeal growths ; the latter, however, are diagnosed by being thoroughly blended with the surrounding tissues, by being very frequently ulcerated, and by the constitutional history and symptoms of the patient. In these cases, should particles be expectorated, or removed during life, with the aid of the laryngoscope, the microscope cannot be relied on for differential diagnosis.

---

[1] In the case of a patient kindly brought under my notice by Mr. Erasmus Wilson in 1865, though there were numerous shining tubercles on the face and other parts, in the larynx there was only thickening of the epiglottis.

Several cases have come under my notice where the histological features were decidedly those of cancer, whilst the clinical history was of a totally opposite character, and *vice versâ.*

Outgrowths,[1] whether of cartilaginous or fibrous character, are not likely to lead to mistaken diagnosis. It is true that the symptoms are often similar, but when the laryngoscope is used, the entire absence of demarcation between the protuberance and the normal tissues, is at once evident. When seen with the laryngeal mirror, they appear rather as swellings or infiltrations, (though of course present no signs of hyperæmia, nor disposition to degeneration or decay,) than as defined tumours. A case of this sort is contained in my Jacksonian Prize Essay,[2] in which the outgrowth was probably of fibrous character, and Virchow's remarks on cartilaginous outgrowths will be found in the section on Pathology of this treatise.

[1] Strictly speaking, of course, the greater number of laryngeal growths are "outgrowths," that is to say, "they are connected with the adjacent parts by continuity of similar tissue, and thus are growths, not in, but of, the parts." The distinctions, however, between growths, or "discontinuous hypertrophies," and "outgrowths," as Mr. Paget has well pointed out, vanish in certain positions, and in the larynx, though nearly all growths are outgrowths, as a matter of convenience, the term is limited in this article to those cases in which the tumours spring from the deeper tissues.

[2] MSS. and Coloured Drawing in the Library of the Royal College of Surgeons.

SECTION V.

—◦◦—

PATHOLOGY.[1]

General Remarks on Pathology.—The investigation into the histology of laryngeal tumours is attended with considerable difficulty, for they are frequently removed in fragments, and even when the whole growth is taken away at a single operation, its origin is often so much lacerated, that it is impossible to tell whether such a neoplasm is a mere out-growth or whether it is a true tumour. Again, inasmuch as these growths are often removed piecemeal, an opinion as to the pathological nature of the growth is apt to be formed after the examination of any one fragment removed.

Virchow[2] has already called attention to the frequent mistakes which occur from the examination of small portions of tumours, even when removed in their entirety, on account of their structure differing so much in different parts; and it will be at once evident that such sources of error are exceedingly likely to arise when a growth is removed in several particles at different times. The various growths found in the larynx appear to be due to perverted development of either the connective tissue or its superjacent structure (the epithelium), or of the parts contained in the connective tissue (the glandulæ and vessels). The *fibro-*

[1] The situation, size, and external appearance of the various kinds of laryngeal growths having been described under the head of Symptoms (Section III., Laryngoscopic Signs), it has not been thought necessary to repeat them in this chapter. It may here be observed, that Appendix D, containing the tabular statement of all cases treated by other practitioners, has been of little use in forming an estimate as to the frequency and situation of different kinds of growths, owing to the information on these points being very incomplete.

[2] *Op. cit.*, vol. i. p. 348.

cellular growths constitute perhaps the type of the connective-tissue tumours, whilst *fibromata* represent those in which the fibrous tissue predominates; *myxomta*, those in which the mucous or embryonic matter is most abundant; *fasciculated sarcomata*, those in which " the embryonic tissue has already undergone a trace of organization and evolution in the direction of connective tissue ;" and *lipomata*, those in which fatty matter is very abundant.

The epithelium being more exposed, is even more subject to perverted development than the connective tissue ; we have, therefore, *simple epithelial growths ;* or when, as is commonly the case, they assume a papillary structure, *papillomata* are formed. From the abnormal evolution of the glandular elements, we have *cystic growths* and *adenomata*, whilst the morbid production of blood-vessels gives rise to *vascular growths*, or *angeiomata*.

Papillomata are by far the most frequent of all the benign growths in the larynx. In my 100 tabulated cases, 67 were judged to be of this character.[1]

These growths occur at an earlier period of life than the other kinds of tumours, nearly all cases found in the first decennial period being either papillomatous or benign epithelial. In the cases of congenital growth collected by Dr. Causit,[2] and in those now reported by myself, the neo-plasms have, when examined, generally been found to be of a warty character. The (supposed) congenital tumours in the Museums of St. Bartholomew's[3] and St. Thomas's Hopitals[4] are likewise of a warty structure.

The rate of growth in this class of tumours varies greatly, but it is generally most rapid at the inception. In one

---

[1] A microscopic examination was made in 29 instances. In Appendices A and B, where the microscope was not used, and the nature of the growth was merely inferred, the pathological character is bracketed at the heading of each case ; thus, " (Papillomatous) Growth on Right Vocal Cord," &c., means that the growth was inferred to be papillomatous ; but where the adjective is written without brackets, a microscopical examination will be found accompanying the description of the case.

[2] *Op. cit.*   [3] Series XXV. No. 17.   [4] Series IV. No. 52.

instance,[1] two growths, placed symmetrically on the posterior
part of the vocal cords, attained the size of split peas in less
than three months; and in another[2] the growth reached the
size of a raspberry in less than nine months.

Papillomata are classified by some pathologists, and
notably by Rokitansky and Virchow, under the head of
Fibromata; but inasmuch as the fibrous element is generally
very sparingly developed, and frequently not to be dis-
cerned, this arrangement does not appear to me to be
justifiable.

Mr. Paget[3] observes, that these growths "may be occa-
sioned either by an hypertrophy of normal papillæ, or may
be entirely new formations in the part. In their general
form and arrangement they have many points of resemblance, but on an enlarged scale, to the papillæ which, in
various localities, constitute natural projections from free
surfaces; more especially from the skin and mucous
membranes. To some extent these papillary growths, in
whatever locality they may be found, correspond in structure
with each other. Their basis substance is formed of connective tissue, which is continuous with that which normally
exists in the part; whilst the free surface is covered by an
epithelium, which may vary in its thickness, and in the number
of its layers, according to the seat of the tumour. Blood-
vessels, and even nerves, enter into the interior of the
papillæ. When a number of these new papillary growths
become aggregated together, they may form a tumour of some
size. Of the *cutaneous* papillary growths, the best-known
example is the common wart, which in many persons forms
in such numbers on the skin of the hands. These warts
consist in an excessive development in length and thickness
of both the dermal and epidermal structures which constitute the papillæ of the skin. The condylomatous growths
which sometimes form in the region of the prepuce, and
about the labia and anus, are excessive developments of

---

[1] Appendix A, Case 80.   [2] Appendix A, Case 86.

[3] *Lectures on Surgical Pathology*, third edition, 1870, edited by Professor
Turner, M.B., p. 591.

the same character; and when they exhibit a very irregular, subdivided surface, they present the well-known cauliflower appearance . . . . Various of the *mucous membranes* are also liable to be affected with abnormal papillary growths. Many of the mucous polypi are complicated with papillary formations. Sometimes the new-formed papillæ are scattered irregularly over a considerable tract of the mucous surface, so as to give it a villous, velvety appearance, though at others they are aggregated into the form of a distinct tumour . . . . The papillæ are recognizable to the naked eye, and impart a distinct villous appearance to the mucous surface, from which they spring. They divide and give off lateral sprouts or branches; they are vascular, and the epithelium which invests them usually corresponds in form with the normal epithelium of the part." This last observation does not hold good with regard to the larynx, for the laryngeal neoplasms have generally a tesselated epithelium, even when removed from parts possessing a ciliated epithelium. My experience on this point is quite in accordance with the observation of Cornil.[1]

The varied character of the component parts of papillary growths, as well as the various stages of their development and degeneration, were seen in many of the specimens kindly examined for me by Dr. Andrew Clark. One case is described by him as "consisting of more or less perfect connective tissue clothed with many layers of epithelium." In another "enlarged racemose glands were found, the terminal vesicles of which were filled with minute nucleated cells and granular matter." In one case, "the growth was found to consist of two sets of particles; one membranous, the others warty or obscurely papilliform. The membranous portion consisted of from twenty to thirty layers of scaly epithelium, surrounded and penetrated by a confervoid growth. The epithelial cells composing the layers were polygonal, flattened, nucleated, and easily affected by weak alkalis and acids. The nucleus of each

---

[1] See *Benign Epithelial Growths*, p. 46.

cell was oval, abruptly defined, rather large in proportion to the containing cell, in most cases surrounded by a clear halo, and in some showing signs of division. The papillary portions consisted of simple outgrowths of nucleated connective tissue, and rudely-formed blood-vessels, clothed with numerous layers of scaly epithelium, similar to those already described. Some of the papillæ exhibited large vacuoles, or spaces filled with colloid matter, which, in one or two instances, had burst through the covering epithelium." The microscopic appearance of one of these papillæ is well shown in Plate I. fig. 1.

Foerster[1] observes that "the papillary blood-vessel is generally broader than the broadest normal capillary ; it is indeed often of colossal size, though its structure is precisely similar to the normal vessel. In many cases, however, the nuclei of the walls are extremely scarce, so that they may, indeed, be altogether overlooked."

Virchow remarks,[2] "that the papillary formation is not merely an hypertrophy or an excess of normal papillary formation, the pathological papilla being derived from a preexisting physiological one, but that every free surface can of itself independently develop papillæ, even in situations where no papillæ previously existed." As regards the mode of development of papillæ, he adds, "that the superficial tissues generate, through exuberance, a certain mass, which, as a rule, appears at first as a small round bud, or a small flat elevation of the free surface. As I found a long time ago, after investigations of the external integument and the tunica albuginea of the ovary, the first outgrowths are very small, amorphous, granular, or homogeneous buds, in which cells become evident at a later period. These gradually increase by multiplication of the cells, and by degrees, they develop into large papillæ or villi. That which happens on a flat surface can likewise occur in a pre-existing papilla. The papillæ can themselves bear

---

[1] Foerster, *Handbuch der Allgemein. Pathol. Anatomie.* Leipzig, 1854, p. 206.   [2] *Op. cit.,* p. 334.

buds, which may also increase ; and thus it may come to pass, in the end, that branch-papillæ are formed. . . . . The whole process has the greatest resemblance to that which regularly takes place, on the free surface of the chorion in the human and mammaliferous ovum, and leads to the formation of the fœtal placenta. The villous portion of the chorion is the physiological example of the papillary hyperplasia, for, to a certain extent, the fœtal placenta may be regarded as a large papillary tumour ; and those new formations are analogous, which, corresponding to this description, are formed under morbid conditions on the free surfaces. . . . Many pathologists believe that, in the formation of every warty tumour, the outgrowth of the superficial vessels is the essential element ; in other words, that the capillaries of the skin, and especially those of the papillæ, widen and lengthen themselves, and gradually push the parts further out. This is decidedly incorrect : equally incorrect for the pathological papillary formation and for that of the villi of the chorion. For if, at all a careful examination is made, it is invariably found that there is a formation of connective tissue, and frequently also of a considerable quantity of epidermic structure, before there is any appearance of vessels, and that vessels are only formed at a later period. . . . . Wherever growth has taken place, an increased formation of granules and cells is found ; and the sprouting can often be seen taking place at the extreme points (of the papillæ), and it may be observed, that whilst the elements at the base are widely separated, the apex is entirely formed of cells, as I have already proved in the case of the villi of the chorion. . . . . In some cases the vessels constitute a very considerable bulk of the growth, and it is then often difficult to distinguish the small portion of connective tissue, which forms a kind of investing membrane. It sometimes, indeed, happens that the vascular supply is so considerable that it extends throughout, even to the free surface, and the whole outgrowth appears like a vascular plexus. If the growth is enveloped with epithelium or epidermis, the epithelial cells are so closely·

applied, that if a single loop be examined, it often appears, as if the epithelium were directly in contact with the capillary wall. When, however, the parts are specially considered, and their mode of development is studied, it is evident that a fine stratum or a kind of adventitious membrane of connective tissue always exists, and that the supposed sprouting out of vessels into the epithelium does not actually occur. If the epithelium be very thin, as it is on mucous membranes, these very broad and thin vessels may, as one can readily understand, be exceedingly exposed."

In his desire to include a considerable variety of growths under the head of Fibromata, Virchow[1] appears to me to have somewhat stretched a point with reference to the papillary tumours. These growths are certainly epidermal or epithelial productions, the areolar tissue being a very unimportant element in their structure. Nevertheless, he remarks that " these growths, whether they contain more or less connective tissue, are in fact of the nature of connective tissue, and must be regarded as outgrowths of pre-existing connective tissue. This character is so apparent, that for a long time they have been described under the name of vegetations. Latterly, an especial value has been attached to the papillary form of growth, and, following Kramer, they have been described under the name of Papillomata. This, however, is quite superfluous, as there are already descriptions enough for individual forms ; it is, moreover, inaccurate, as the nature of the tumour is essentially that of connective tissue in papillary form. The generic name must therefore be Fibroma, and the term ' papillary' can only be applied qualitatively."

Papillomata show a certain disposition to recurrence, the disease having reappeared in 4 instances out of the 67 cases treated by me. In two others, where the neoplasm was not entirely eradicated, there was rather rapid and considerable increase of the remnant. I have seen many of my

---

[1] Virchow, whilst including papillary growths under the head of Fibromata, nevertheless admits that there are some neoplasms (described by him under the term of Acuminated Condylomata) in which "the growth might almost be correctly reckoned amongst epidermoidal tumours."

patients at long intervals after they have been cured, but, of course, there remain some who were lost sight of at a comparatively early period. The proportion of the recurrence is therefore, in all probability, rather greater than my statistics indicate.

Relation of Papillomata to Warty Cancer.—Many kinds of neoplasm undergo papillary development ; but it is only epitheliomata that can give rise to difficulty in deciding as to the nature of a growth.

Mr. Paget remarks,[1] that, in warty cancer, " a certain portion of the skin or mucous membrane is infiltrated with epithelial cancer-structures ; on this, as on a base more or less elevated and imbedded, the papillæ, variously changed in shape, size, and grouping, are also cancerous ; their natural structures, if we except their blood-vessels, which appear enlarged, are replaced by epithelial cancer-cells. And herein is the essential distinction between a simple or common warty or papillary growth, and a cancerous one, or warty cancer. In the former the papillæ retain their natural structures ; however much they may be multiplied or changed in shape and size, they are either merely hypertrophied, or are infiltrated with organized inflammatory products ; however abundant the epidermis or epithelium may be, it only covers and ensheathes them. But in the warty cancer the papillæ are themselves cancerous ; more or less of their natural shape, or of the manner of their increase, may be traced : but their natural structures are replaced by cancer-structures ; the cells, like those of epithelium, lie not only over, but within, them." Although the foregoing description is of much value, I do not think that it is sufficiently precise to enable the microscopist to differentiate accurately between the simple warty, and truly cancerous, neoplasms. The relation of cells is so much a matter of accident, and their form and contents so dependent on their period of development, that the " cancer-structures " of epithelioma do not form a reliable basis for differentia-

[1] *Op. cit.,* p. 706.

tion. It appears to me that the "epidermic globes" of Lebert, or "laminated capsules" of Mr. Paget (Plate I. fig. 8), alone furnish the crucial test. These, though present in the epidermic accumulations of cysts,[1] are, I believe, never found in simple papillary growths, and are "well marked in nearly every epithelial cancer."

Benign Epithelial Growths constitute a small proportion of laryngeal neoplasms. In these tumours, the epithelial scales do not clothe papillæ, but form continuous layers of more or less undulating character. Nearly all laryngeal growths have an epithelial covering; but there are some neoplasms in which the whole growth is made up of epithelial cells (Plate I. fig. 2). These constitute, in fact, simple hypertrophy of the normal epithelium of the larynx: as already observed, many of the cases of (supposed) congenital growth are of this simple structure. In nearly all cases of laryngeal growth, the epithelium is of the tesselated variety; and Cornil[2] remarks that this holds good "even when the growth is developed in a part of the larynx, in which the mucous membrane possesses cylindrical epithelium." Ciliated epithelium has, however, been found on the surface of a laryngeal neoplasm in three cases.[3] In my 100 tabulated cases, benign growths were present in 5 instances; of these two occurred in young children; and one other instance[4] is related amongst my untreated cases. These growths, from the great diversity of the shape of the cells, are, when examined with the microscope, occasionally mistaken for cancer, but they never contain laminated capsules, and can therefore be easily distinguished from true epithelioma.

Fibromata, though not nearly so common as papillomata, are next in order of frequency to those neoplasms. They

---

[1] Mr. Paget, *Op. cit.*, p. 720. See also Virchow, *Archiv*, vi. p. 200, quoted by Mr. Paget.

[2] Quoted by Krishaber (*Op. cit.*) from Dr. Cornil's unpublished MSS.

[3] Ehrmann (*Op. cit.*, Observation xxii.); Follin (Appendix D, Case 56); Krishaber (Appendix D, Case 187).

[4] Appendix B, Case 9.

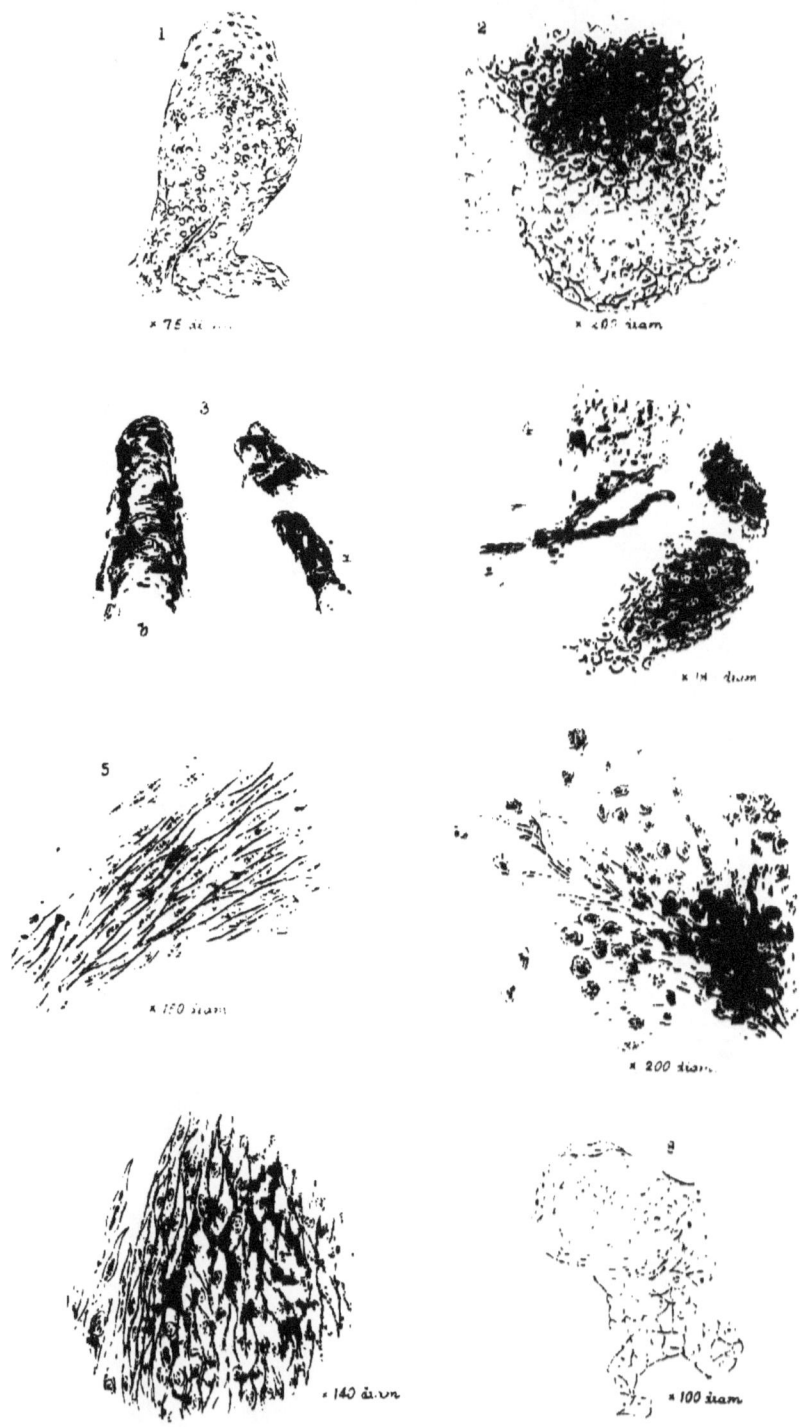

DESCRIPTION OF PLATE I.

––—

Fig. 1 represents an enlarged Papilla taken from the growth removed in Case 82, and is a type of these formations.

Fig. 2 illustrates the section of the Benign Epithelial Growth removed in Case 98.

Fig. 3 represents—(a) fragment of a Fibroma, Case 78—(b) The same fragment slightly enlarged.

Fig. 4 represents a section of a Fibro-cellular Growth, portions of which are of the character of a Myxoma; taken from Case 99.

Fig. 5 illustrates a section of Fibro-epithelial Growth from Case 74.

Figs. 6 and 7 represent varieties of Fasciculated Sarcoma, from Cases 59 and 49. No. 6 illustrates a case of ordinary Fasciculated Sarcoma or Recurrent Fibroid; and No. 7, a case of Fibro-nucleated Growth.

Fig. 8 represents a laminated capsule or nested cell removed in Case 87.

To come in at pp. 46-47.]

were found to exist in 11 per cent. of my cases.[1] The youngest patient affected was 27 years of age, the eldest 57. In this class of neoplasm, the rate of growth is much slower than in that of papillomata.

Though generally situated in the submucous tissue, fibromata are supposed to grow in some cases from the perichondrium;[2] when examined microscopically, they are seen to consist of bundles of white fibres, diverging and interlacing in various directions (Plate I. fig. 3), and are generally covered with several epithelial layers. Krishaber,[3] indeed, observes that this epithelial covering is invariably present. I have, however, examined several cases microscopically, which possessed no epithelial envelope, and in which there was no evidence of any desquamation having previously taken place.

Mr. Paget[4] remarks "that, in the examination of sections, the most usual characters that one sees, are that the tumours present a greyish basis-substance, nearly homogeneous, and intersected with opaque, pure white bands and lines. They have a general resemblance in their aspect to a section of fibro-cartilage, such as that of the semilunar or the intervertebral cartilages. Many varieties, however, appear; the basis-substance tending towards yellow, brown, or blue, and the white lines being variously arranged." Occasionally the homogeneous basis-substance has not undergone fibrillation, and we then find "amorphous masses of a coagulated proteine compound." This was exemplified by a case[5] kindly examined for me by Dr. Andrew Clark.

These growths show no disposition to recurrence.

Fibro-cellular Growths, or Mucous Polypi, consist of more or less perfectly developed fibro-cellular tissue, and have

[1] A microscopic examination was made in every instance but one, in which the growth was inferred to be a fibroma from its smooth surface and compact, white, internal structure.

[2] *Handbuch der speciel. pathol. Anatomie,* von Dr. August Foerster. Leipzig, 1854.

[3] *Op. cit.,* p. 752. [4] *Op. cit.,* p. 476.

[5] Appendix A, Case No. 7.

diffused through their substance a greater or less quantity of serous-like fluid. They are comparatively rare in the larynx, being found in only 5 per cent. of the cases here reported. The ages of the patients were 18, 21, 28, 30, and 65.

In the few cases that have been seen, the rate of growth appears to have been rather slow.

As regards the microscopical structure of these polypi, Mr. Paget[1] remarks "that they consist of delicate fibro-cellular tissue, in fine undulating and interlacing bundles of filaments. In the interstitial liquid, or half-liquid substance, nucleated cells appear, imbedded in a clear or dimly granular substance; and these cells may be spherical or elongated, or stellate; imitating all the forms of such as occur in the natural embryonic fibro-cellular tissue; or, the mass may be more completely formed of fibro-cellular tissue, in which, on adding acetic acid, abundant nuclei appear. In general the firmer the polypus is, the more perfect, as well as the more abundant, is the fibro-cellular tissue." The tuberous fibromata of Virchow correspond, to some extent, with the fibro-cellular polypi of Paget.

Though mucous polypi of the nose have a great disposition to recurrence, or to develop in a multiple form, the same fortunately cannot be said of these growths in the larynx. When removed, they have no disposition to recur. In each case, also, there has been only one growth.

Myxomata, or true mucous growths, are exceedingly rare in the larynx, and I have not myself met with a single instance, in which a laryngeal neoplasm was entirely of a myxomatous nature. In one case,[2] however, a small portion of a large growth, mainly consisting of fibro-cellular tissue, approached the true myxoma of Virchow. The portion referred to was about the size of a large currant, of jelly-like consistence, and contained a few oval and elongated corpuscles, and an obscurely fibrillated intercellular substance (Plate I. fig. 4). The only other instance that I know is that of Professor

---

[1] *Op. cit.*, p. 456.   [2] Appendix A, Case 99.

Bruns.[1] In his case, the patient was a man aged 56, and the symptoms had been coming on for about eight years. The specimen was reported by Dr. Schüppel to be a *myxoma hyalinum*. It consisted of a greyish-yellow, transparent, firm, gelatinous substance, arranged in more or less complete lobes or lobules, by bundles of connective tissue, which passed through it, in various directions. The gelatinous mass consisted of an abundant perfectly clear homogeneous intercellular substance, with a few scattered cells of various sizes and shapes, all provided with thin thread-like prolongations. The growth was contained in a fibro-cellular envelope, and covered with about eight layers of epithelium.

Lipomata, or fatty tumours, are rarely found in the larynx, only one case, which occurred in the practice of Professor Bruns,[2] having hitherto been published. Virchow has called attention to the fact, that both lipomata and myxomata originate, by preference, in those parts in which a cellulo-adipose tissue exists, and the extreme scarcity of this compound tissue in the larynx, accounts for the great rarity of both these varieties of growth in that part. In Bruns' case, the patient was 25 years old, and the growth was believed by the distinguished professor to have been congenital. The lipoma was very large, and entirely occluded the orifice of the larynx; it extended between the ary-epiglottic folds laterally, and from the epiglottis in front, to the arytenoid cartilages posteriorly, and appears to have been attached to the mucous membrane over the left arytenoid cartilage, and to the left ary-epiglottic fold. The greater part of it was removed in fifteen galvano-caustic *séances.*

A careful microscopic examination was made by Professor Schüppel.[3] The investing membrane of the growth consisted of a stellate-fibred connective tissue, intermingled with tolerably abundant, delicate, elastic fibres; it was unusually rich in thin-walled blood-vessels, and presented a smooth surface, free from papillæ. The epithelium was of

---

[1] *Polypen des Kehlkopfes,* by Professor Von Bruns." Tübingen, 1868, p. 17.
[2] *Op. cit.*   [3] *Op. cit.,* p. 91.

a laminated character, consisting of about fifteen layers. The membranous envelope contained two oval fatty tumours of about the same size. The fat-cells were of medium size and partially filled with margarine. Between the little clusters of fat, bundles of fibrous tissues were seen, which were evident, to the naked eye, as opaque, whitish-yellow, curved, lines. The neoplasm also contained a small ˊcartilaginous growth about the size of a hemp-seed, surrounded on all sides by connective tissue.

Fasciculated Sarcomata (*Synonyms:* Spindle-celled Sarcoma; Fibro-plastic Tumour of Lebert; Recurrent Fibroid of Paget) constitute a variety of growth, which is comparately unfrequent in the larynx, only three cases [1] having come under my notice, and there being but six amongst the cases treated by other practitioners (Appendix D).

One of my cases occurred in a man aged 42, the others in women, aged respectively 53 and 43. In the last case dysphonia had existed for 23 years, so that the tumour must have grown very slowly, or its growth must have been entirely arrested after a time.

On examining these growths with the microscope, two of them showed the characteristic appearance; viz., long fusiform cells, arranged in such a way that the tapering extremities of one set of cells were in contact with the expanded portions of the next set. This appearance is shown in Plate I. fig. 5, and in the woodcut, fig. 84, Case 95. In the other case the growth presented rather the appearance of a *fibro-nucleated* tumour, the nuclei being very distinct. The appearance is shown in Plate I. fig. 7. Mr. Paget [2] makes a separate class of these fibro-nucleated growths, which he considers "occupy a kind of middle ground between innocent and malignant tumours." He remarks, moreover, "that it would be wrong to endeavour to draw many conclusions from so small an experience as yet exists of these tumours."

It is scarcely necessary to observe, that the fasciculated

----

[1] Appendix A. Nos. 59, 49, and 95.   [2] *Op. cit.*, p. 605, *et seq.*

sarcomata have received the name of recurrent fibroid, from their disposition to recur. In one of my cases [1] the recurrence is continuous ; in another [2] the recurrence took place after an interval of eighteen months ; and in the third [3] recurrence has not yet taken place.

Cystic Tumours are comparatively rare. Though the minute glands of the mucous membrane of the larynx often undergo slight cystic enlargement, they seldom attain such a magnitude as to render operative procedure necessary. Of my 100 tabulated cases, only two were of the true cystic character. A case of this sort was successfully operated upon by Mr. Durham [4] in the year 1863, and four others will be found in the cases collected from all sources.[5] One of my cases occurred in a woman, aged 44, the other in a young man 22 years old. I am not aware that any case has been observed before the age of 11 years, which was the age of Mr. Durham's patient. In the four other published cases,[6] the ages of the patients were 38, 64, 50, and 15. In my second case, the development of the cyst, which was of large size, and apparent on both the upper and under surfaces of the epiglottis, had been very slow ; whilst in the first case, it caused considerable inconvenience six months after the first symptom of its presence had been noticed. These cysts are situated in the sub-mucous tissue : they generally have dense walls, and are more or less completely filled with thick, white, semi-fluid, sebaceous-like material ; sometimes, however, the product is a thin yellowish or brown fluid. On microscopic examination, these contents are seen to consist of epithelial cells undergoing fatty degeneration. In the thinner fluids, some of the proper secretion of the glandulæ still exists, in combination with epithelial scales, granular fatty matter, cells, and molecules. One case,[7] however, has been placed on record by Dr. Johnson, to which

---

[1] Appendix A, Case 59. [2] Appendix A, Case 49.
[3] Appendix A, Case 95.
[4] *Transactions of the Medico-Chirurgical Society,* 1863, and Appendix D, Case 51. [5] Appendix D.
[6] Appendix D, Cases 57, 84, 99, and 129. [7] Appendix D, Case 99.

the foregoing description does not apply, the wall of the cyst being thin and membranous, and the contents sero-sanguineous.

Although, from our knowledge of other retention cysts, we might have anticipated that cystic tumours of the larynx would be likely to fill again, experience, as far as it goes, seems to show that when these laryngeal cysts have been thoroughly laid open, their contents emptied, and the cyst-wall cauterized, there is no tendency to recurrence.

Adenomata, or glandular tumours, are seldom met with in the larynx, though acinous gland-structure is often found in papillary growths;[1] occasionally, however, the entire neoplasm consists of an hypertrophied racemose gland (fig. 70, Case 79). Two instances of this sort have come under my notice.[2] In the first, the growth was situated below the anterior insertion of the vocal cords; in the second, the epiglottis was the seat of the disease, and here the growth attained the size of a large cherry in the course of three months. This specimen was exhibited at the Pathological Society,[3] and was pronounced, after being carefully examined by two eminent microscopists, to be a case of "adenoid cancer." As, however, after a year's interval, there has been no recurrence, it is probable, that the growth was a simple adenoma. One case has also been reported by Bruns,[4] occurring in a man aged 74. In a case published by Drs. Herard and Cornil,[5] "the culs-de-sac and canals of the hypertrophied glands were lined with cylindrical epithelium. Between the glandular culs-de-sac, very fine areolar tissue was found, traversed by numerous capillary vessels; and in the areolar tissue there were a great number of granules, lymphoid cells, and fusiform corpuscles."

---

[1] The reverse of this is stated by Drs. Cornil and Ranvier in their useful little *Manuel d'Histologie pathologique*, p. 289; but Dr. Andrew Clark has repeatedly found portions of racemose glands in the growths I have removed.

[2] Appendix A, Cases 79 and 88.

[3] *Transactions of the Pathological Society*, vol. xxi.

[4] *Polypen des Kehlkopfes.* Tübingen, 1868, p. 30; and Appendix B, Case 123.

[5] *Phthisie*, p. 92, quoted by Krishaber (*Op. cit.*).

Angeiomata, or vascular tumours, are exceedingly rare in the larynx, and I know of no other case than that now reported.[1] In this instance the neoplasm, whilst *in situ*, was bluish-black, but after its removal it was pinkish-red. On microscopic examination, its structure was seen to be obscurely fibrous, the fibres being very closely intermingled, and the vessels so closely and tightly packed together, that they could not be isolated. No person who saw the growth, before its removal, could doubt its being a true angeioma, and in judging of its nature, I rely more on the appearances presented by the living growth, than on the subsequent microscopic examination. The specimen seemed to belong to the Venous Vascular Tumours of Mr. Paget—the Angeioma Cavernosum of Plenck. In the case referred to, the growth had only given rise to uneasiness during the previous six months. From the situation, however, in which it grew, it might have existed for many years previously, and indeed it may even have been congenital, and may have only caused trouble, when it became congested.

From the frequent congenital occurrence of these vascular growths in other parts, one might have expected to have found them in the larynx at birth. As far, however, as investigations have at present been carried, the congenital growths in the larynx appear to be either simple benign epithelial productions, or papillomata.

There is no evidence as to the tendency to recurrence ; but when such growths are destroyed in other parts of the body, they show no disposition to return.

Compound Growths are not unfrequent ; indeed, it is often exceedingly difficult to determine to which class of neoplasms a given growth belongs. In one variety indeed the compound elements are so constantly present, that they constitute a subdivision, viz., that of fibro-cellular growths. In my 100 tabulated cases, I have classed each tumour under the head of its predominating constituent. In one of the cases,[2] however, the amount of fibrous tissue and epithelial cells was so nicely balanced, that I have felt bound to

[1] Appendix A, Case 89. [2] Appendix A, Case 75.

describe it as a fibro-epithelial growth (Plate I. fig. 6) ; and in another case,[1] described as fibro-cellular, a portion of the growth was of distinctly myxomatous character.

Other kinds of Growth.—It is as well to remark that hydatids are stated to have been found in the larynx.[2] Ryland[3] remarks, that "a case of this sort developed in one of the ventricles of the larynx, has been known to project so far into the cavity of this organ, as to give rise to all the symptoms which usually attend a foreign body there." On this subject, Foerster observes,[4] that "*mucous polypi* were described, as hydatids, by the older authors." Ryland also refers to cases of cartilaginous tumours of the larynx ; but the examination of these growths was made at a period (1835) when histology was quite in its infancy, and the account, therefore, is not of much value. Rokitansky does not mention the occurrence of cartilaginous tumours in the larynx, but Virchow,[5] limiting the term of Enchondroma to heterologous growths, describes those cartilaginous tumours, which arise in connection with pre-existing cartilage, as Ecchondroses. He especially calls attention to the occurrence of these growths in the larynx, and remarks that, " whether arising from the thyroid or cricoid cartilage, they generally grew towards the cavity of the larynx." This is not, however, invariably the case, for in a specimen which I exhibited last year at the Pathological Society,[6] a growth about the size of a bantam's egg, originating from the cricoid cartilage, extended downwards and forwards in front of the trachea. " The cartilaginous outgrowths," says Virchow, " are some- times broad and flat, sometimes circumscribed and nodular. On examining the larynx (with the laryngoscope), an out- growth of this sort, as it has an epithelial covering, is easily

---

[1] Appendix A, Case 99.
[2] Andral, *Anat. Pathol.*, Translation, vol. ii. p. 459.
[3] Ryland, *Diseases of the Larynx*, p. 226.
[4] Foerster, *Handbuch der speciellen pathol. Anatomie.* Leipzig, 1854, p. 216.
[5] *Op. cit.*, p. 438, *et seq.*
[6] *Transactions of the Pathological Society*, vol. xxi. p. 58.

mistaken for a polypus, and at the present time, when laryngeal growths are studied with so much interest, these cases deserve special notice, as, from their thickness and hardness, any operation, carried out *per vias naturales* is altogether impossible."

Degeneration of Growths.—The laryngeal neoplasms exhibit very little tendency to retrogressive changes. Occasionally, but very rarely, papillomata undergo fatty degeneration, and probably in those few cases in which spontaneous expulsion of the growth has taken place, this change had previously occurred. Caustics may also perhaps, in some cases, promote these degenerative evolutions. In one case of fasciculated sarcoma, which is still under my care, portions of the growth appear, from time to time, to undergo saponification.

COMPARATIVE PATHOLOGY.

As early as the year 1829 Albers[1] reported a case of polypus in the larynx of a cow, and since that time growths have been found in the larynx of the same animal, as well as in the horse and the dog. The museum of Dresden contains two specimens of growths in the horse ; and there is one also in the museum of the Royal Veterinary College of London. This last one is situated on the epiglottis, and appears to be of cystic character. The museums of Dresden and Fribourg[2] each contain an example of tumours in the larynx of the cow.

I recently exhibited at the Pathological Society a specimen of this disease in the larynx of a dog.[3] Several neoplasms

---

[1] *Pathologie und Therapie der Kehlkopfskrankheiten*, p. 204.

[2] The growths in the museums of Dresden and Fribourg here referred to, are described and figured by Ehrmann, *Op. cit.*, pp. 29, 30, and Plate VI.

[3] The following is the history of the case :—" Dash, æt. 3½ years, a cross-breed of lively disposition, enjoyed good health till the spring of 1870, when it was noticed that his bark was weak. The bark grew more feeble in the summer, and in the autumn his respiration became difficult, especially when he underwent exertion or was excited, or when he passed from a warm into a cold atmosphere. In these attacks his inspiration was stridulous, as in child-crowing ; and

were found completely blocking up the subglottic region :
one of these, the size of a small bean, was situated at
the posterior part of the left vocal cord, and a smaller
one on the same part of the right vocal cord ; there was
also a fringe of very minute growths along the ventricular bands (Plate V. fig. 3). On microscopical examination,
the growths were found to consist of minute granular cells—
probably epithelial cells, undergoing degenerative metamor-
phosis.

This case is, as far as I am aware, at present, unique ; but
as dogs often suffer from dysphonia and dyspnœa, I have no
doubt, that if these growths were searched for, they would be
frequently found.

he always lay flat on his belly, with his head stretched straight out, resting on
the floor. He became much thinner and weaker towards the end of the year ;
and in December a veterinary surgeon, who was called in, said that 'the lungs
were so extensively diseased that little remained of them.' As the case was hope-
less, he prescribed hydrocyanic acid. *Post-mortem appearances*—Lungs perfectly
healthy, but a mass of growths was seen blocking up the larynx."

As this interesting case reflects somewhat unfavourably on the veterinary art,
it is only just to observe that in two of the cases occurring in animals, referred to
by Ehrmann, the situation of the obstruction was accurately diagnosed during life,
and that tracheotomy was successfully performed. It may also be added, that, at
the time those animals were operated upon, no case had been recorded in which
tracheotomy had been performed on the human subject on account of a laryngeal
growth.

SECTION VI.

PROGNOSIS.

THE tendency to death being by suffocation, and the most common symptom caused by a growth in the larynx being dysphonia, the prognosis has to be considered in relation to these two circumstances. In the few cases in which dysphagia is present, the neoplasm is generally attached to the epiglottis, and can therefore be easily removed. Under these circumstances a favourable prognosis may be given.

In relation to Life.—Growths in the larynx which cannot be removed with the aid of the laryngoscope are always attended with danger to life, which is either immediate or remote, according as the neoplasm is large or small. The gravity of the prognosis is also affected by the age of the patient, the disease being, *cæteris paribus*, less dangerous in the case of adults than young children.

In *adults* death is not likely to take place from suffocation, unless the patient refuses to submit to proper treatment. Of course, if tracheotomy is performed, this peril is at once avoided ; but it must not be forgotten that, even in opening the windpipe, there is a very slight, though still an appreciable risk. The disposition to bronchitis, which is often the immediate result of tracheotomy, when prolonged dyspnœa has prevailed, must also be taken into consideration. In nicely balancing the remote effects of tracheotomy, it is also well to remember that, when a canula is permanently worn, there is always a slight risk of its giving rise, at some future time, to disease of the cartilages of the larynx. Even, however, after tracheotomy has been performed, and both the

I

immediate and remote dangers of the operation have been passed through, the presence of a growth in the larynx is not without danger. Suffocation has been warded off, but if the neoplasm continue to grow, dysphagia may come on. Hence, in order to extirpate the growth, it may become necessary to divide the thyroid cartilage, an operation always attended with great danger, as will be seen by reference to the remarks on this subject (page 94).

In *children*, as the larynx is, of course, much smaller, the disposition to spasm is much greater, and not only treatment, but even accurate diagnosis, is much more difficult ; the prognosis, therefore, as regards a fatal termination, is more serious. Out of the 46 cases collected by Causit,[1] a fatal termination occurred 21 times ; and this large proportion results from the fact of his Essay having had particular reference to the occurrence of the disease in infants. It should also be added that many of his cases were treated before the invention of the laryngoscope.

The presence of a growth in children is not unlikely to cause serious congestion of the larynx, and should croup arise, a fatal termination is almost certain. There are two specimens in the Museum of Guy's Hospital,[2] which illustrate the great danger of these associated conditions. In one specimen the larynx contains a small epithelial growth, and the whole of the mucous surface is covered with a false membrane. The other specimen presents "minute cauliflower vegetations on the vocal cords and a thin layer of coagulable lymph, described as 'chronic croup, covering the mucous membrane generally.' The child was about four years of age and had lost its voice for five months."

In children also the prospect in relation to tracheotomy, both as regards the operation itself and its immediate results, is unfavourable. In one of my 100 tabulated cases (Appendix C), a child, aged $2\frac{1}{2}$ years, died after tracheotomy, and one fatal case will be found in Appendix B. In this case the infant was so weak that the

---

[1] *Op. cit.*   [2] Specimen 1690, 95 ; and 1702

operation, which was refused by the mother, could scarcely have been successful.

In relation to Voice.—As regards the voice, a favourable opinion may, as a rule, be given if laryngoscopic treatment can be employed. If the fauces be not abnormally sensitive, if the upper opening of the larynx be of average size, if the growth be single, and if it be pedunculated, there is every probability that the voice will be restored. If the opposite conditions prevail, the prognosis is less favourable. When the growths are sessile, very numerous, and apparently closely incorporated with the subjacent tissues, the prospect of restoring the voice is much more doubtful.

When an exceedingly minute growth is situated on one of the vocal cords, or when the growth is unusually large, the prognosis in relation to voice must be more guarded ; in the first instance, because the difficulty of seizing the growth is very great ; and in the second case, because as the insertion of the growth frequently cannot be determined, the certainty of its complete removal cannot be predicted. In the 89 of my 100 cases in which the voice was impaired (Appendix C) it was completely restored in 70 instances, and was improved in 16. In the 189 cases treated by other practitioners, 109 cases were cured, that is, 57·6 per cent., and improvement took place in 68 cases, that is in 35·9 per cent.[1]

When laryngoscopic treatment cannot be carried out, and the thyroid cartilage has to be divided, the prognosis, as to recovery of the voice, is unfavourable, as will be seen by reference to the analysis of the Thyrotomy Table, page 95.

In giving an opinion as to the ultimate result of these cases, even when treatment is adopted with success, the disposition to recurrence must not be forgotten. In the section on pathology, it may be seen that whilst papillomata show a continual disposition to reproduction, other laryngeal growths, with the exception of fasciculated sarcomata, seldom recur.

---

[1] The result in Table D can only be approximately ascertained with reference to voice, as it is often stated that the case is "cured" or "improved" without any mention of the state of the voice. In some cases, in which "improvement" is stated to have taken place, there certainly was no benefit to the voice (Cases 16, 107, and 112).

SECTION VII.

—⚬—

TREATMENT.

BEFORE considering the subject of treatment, it may be well to observe that there are a few cases in which operative procedure is not required. Thus small growths on the epiglottis, or ventricular bands, which cause little or no inconvenience, may well be left alone. This remark especially applies to fibromata, which grow much less quickly and are frequently arrested in their development. In these cases, all that is necessary, is, to make a periodical examination of the larynx, once every two or three months, to see that the neoplasm does not increase in size. Several cases have come under my observation, during the last three years, in which small warts, after attaining a certain size, have not undergone any further development. In the year 1868 I examined a gentleman's throat and found a small growth, about the size of a pearl barley grain, situated on the left vocal cord. I was not aware that he had been previously examined with the laryngoscope, and I gave him a drawing of the appearance of his larynx. On the next day he brought with him a sketch made by Professor Czermak, in 1863. The growth was found to be identical in each drawing ; it had not undergone the slightest change. In this instance I did not advise treatment.[1]

[1] The patient was a very earnest Member of Parliament, and during those five years had exercised his voice very constantly in the House of Commons, not only as a frequent speaker, but also by vigorously joining the chorus of applause and disapprobation which accompanies the debates. It is all the more remarkable that a growth, actually upon one of the vocal cords, should have remained stationary under such circumstances.

In addition to small growths in unimportant parts, it sometimes happens, that the neoplasm is not sufficiently defined to admit of its removal.[1] In other cases, where, in consequence of the advanced age or occupation of the patient, the voice is of little importance, no treatment need be urged unless the respiration be also affected.[2]

In by far the greater number of cases, however, that come under notice, it is necessary to adopt measures for the removal of the growth, or for the relief of the symptoms it causes, and the treatment may be either palliative or radical.

Palliative Treatment consists in placing the patient, in such a condition, as to relieve him of immediate danger to life. This plan of treatment is called for in all cases, where the growth greatly interferes with respiration, where for any reason laryngoscopic treatment cannot be carried out, and where the patient is unwilling to permit an extra-laryngeal operation.

The only safe palliative treatment consists, of course, in the operation of tracheotomy, and it must be recollected, as already observed, that this operation affords absolute protection only as regards death from suffocation. When growths situated in the cavity of the larynx attain a very large size, they are apt after a time to interfere with deglutition. In such cases, therefore, though tracheotomy may have removed the original source of danger, at a later stage progressive dysphagia may occur.

Radical Treatment may be conducted either *internally*, through the natural upper orifice of the larynx, that is, with the aid of the laryngoscope ; or *externally*, or by direct incision into the larynx ; or by the *combined method*, tracheotomy being first performed, to place the patient in a condition of safety, and the growth being subsequently removed through the mouth.

[1] A brief outline of a case of this sort, which I saw in consultation with Dr. Hyde Salter in 1866, is given in Appendix B (Case 5).
[2] Appendix B, Case 6.

The Removal of Growths by Internal or Laryn- goscopic Treatment.

This method represents, perhaps, the greatest triumph which the laryngoscope has effected. By the removal of a growth in this way, no chance of danger is incurred, little or no pain is felt, and scarcely a drop of blood is lost. By an operation of this simple character, the long-lost function of a most delicate organ may be almost instantly restored, and a morbid condition, threatening the immediate extinction of life, may be at once and for ever removed.

Some cases, indeed, are not so easily cured. The operation may give rise to a slight pricking pain, it may momentarily embarrass respiration, it may cause a slight hæmorrhage from the larynx, and it may require to be frequently re- peated ; but with perseverance on the part of both patient and practitioner, the result is seldom doubtful.

The removal of growths from the larynx requires ingenuity on the part of the operator in overcoming difficulties by means of mechanical contrivances; but above all, perhaps, the in- telligent co-operation of the patient is necessary. Although greater *éclat* is often derived from the removal of a large growth than a small one, it will be readily understood, that, *cæteris paribus*, the smaller the growth the greater the difficulty of its removal. As a rule, a growth of mode- rate dimensions, that is, one between the size of a horse- bean and a Barcelona nut, is most easily seized. Of course, the difficulty partly depends on situation, the posterior portion of the glottis being more accessible than the an- terior, and the upper part of the larynx than the lower. The difficulty is immensely increased when the growth is situated below the vocal cords. It may be remarked that growths, when actually removed from the larynx, are generally much larger than they appear when seen with the laryngoscope. Hence, when only a piece has been removed, one is apt to think that the entire growth has been taken away. For this reason, after the evulsion of what may appear to be the whole growth, it is important not to express an opinion to the patient, that the growth has been completely removed, until by a subsequent laryngoscopic examination, this has been proved to be the case. It is also necessary to observe that operations of any kind on the larynx occasionally give rise to irritation and spasm, and may even cause inflammation. Hence a patient, who was before only hoarse, may become completely aphonic, and one who previously breathed fairly well, may, after the removal of a portion of growth, be affected with considerable dyspnœa, and tracheotomy may be urgently called for. The latter sequel has occurred three times in my practice. It is especially likely to happen when the growth is large, and blocks up the greater part of the laryngeal canal. In these cases it would perhaps be better to perform tracheotomy in the first instance; but, inasmuch as tracheotomy always introduces an element of danger, however slight, and must cause some pain and inconvenience to the patient, the practitioner naturally feels anxious to avoid it. Under these circumstances, he may be content with previously warning the patient of the contingency likely to arise.

Several different Kinds of Instruments, and indeed different modes of treatment, are often required in the same case; and it not unfrequently happens, that, after instrumental treatment has resulted in some benefit, caustics may be subsequently applied with advantage. It is true that certain kinds of instruments are better adapted for certain kinds of growths: thus the short sessile growths—the most common in the larynx—can be most easily removed with forceps; cystic tumours only require incision, and small fibromata may frequently be treated by division of their base. On the other hand, pedunculated growths are favourable to the use of wire-loops, écraseurs, and guillotines. Much depends on habit, and an operator is apt to give undue credit to that instrument which he is most accustomed to use. I think it right to state, however, that, personally, I have found forceps, of various construction, the most useful instruments, and have employed them exclusively in by far the larger number of cases.

Dr. Fauvel, of Paris, who is one of the most skilful and experienced operators, employs forceps almost invariably,

whilst Dr. Tobold[1] prefers *knives.* On this subject he remarks : " I believe, indeed, that I do not go too far when I
assert that a simple strong wire, curved like a catheter, and
terminating at its extremity in a pointed, double-edged

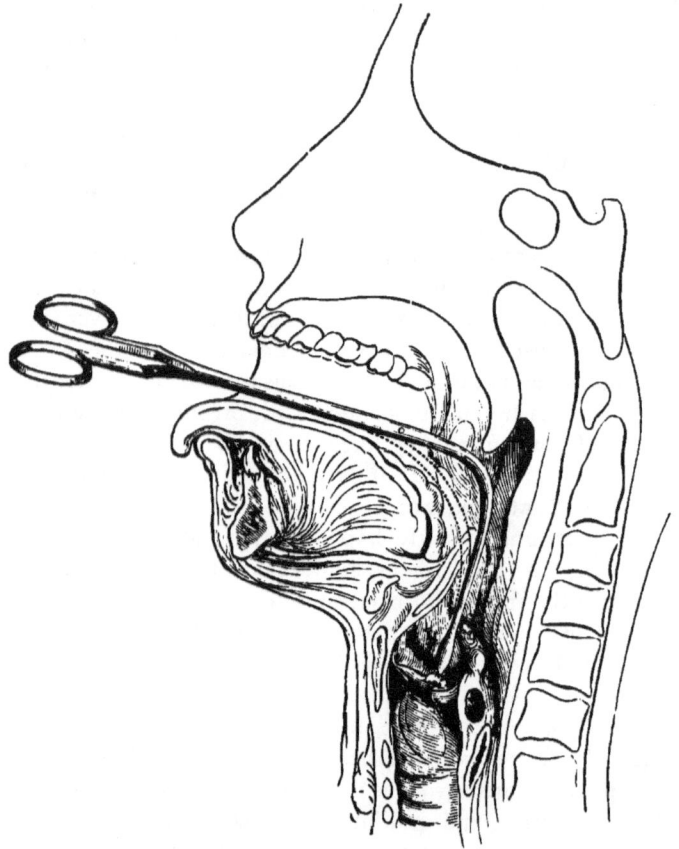

Fig. 2.

knife, is the most convenient instrument for the majority of
cases." This quotation, it will be seen, has reference not
only to the kind, but to the curve of the instrument. Most
practitioners have their laryngeal instruments curved like a
catheter ; but I have long employed those of a more angular

---

[1] *Die chronischen Kehlkopfkrankheiten.* Berlin, 1866, p. 216.

form. In a catheter the two extremities are at right angles to each other ; but the angle is reduced to a minimum by a large curve or sweep. This curve, though well adapted for the urethra, is much less suitable for the larynx ; and if on the other hand, the right angle, slightly smoothed down, is left, the instrument in passing into the larynx is kept free of the epiglottis. My meaning will be at once clear, on reference to fig. 2. It will be seen that both the catheter-curved instrument (indicated by dotted lines), and my rectangular instrument reach the same spot ; but whilst the former touches, and even presses against the epiglottis, the latter avoids it.

Drs. Walker, Bruns, Moura-Bourouillou, and others, have employed écraseurs or wire loops ; and Drs. Johnson and Gibb, who have used them with marked success, appear to have confined themselves entirely to these kinds of instruments.

The following table shows the result of various methods of eradication in those of my cases in which treatment was pursued *per vias naturales.*

	No. of Cases.	Cure.	Improvement.	Negative.	Death.
Caustic Solutions	3	1	2	—	—
Galvanic Cautery	4	2	1	1	—
Tube-forceps	34	25	7	2	—
,, and Caustics	2	2	—	—	—
,, and Common Forceps	9	9	—	—	—
,, and Stoerk's Ecraseur	4	3	1	—	—
Common Laryngeal Forceps	26	23	3	—	—
,, ,, and Crushing	3	1	2	—	—
Wire-loop	1	—	1	—	—
Various Instruments	3	2	1	—	—
Knives { Incision of Cystic Tumours...	2	2	—	—	—
{ ,, of Base of Growth...	2	2	—	—	—
Total ...	93	72	18	3	0

Mr. Durham[1] judiciously remarks on a somewhat analogous table of cases which he has collected, that " the above table must be taken for what it may be worth. It certainly appears to indicate the methods of operating that have

[1] *Op. cit.*, p. 584.

hitherto been found most successful. But it cannot for one moment be regarded as affording any trustworthy guide to the method that ought to be adopted in any particular instance that may come under observation. In deciding upon the course to be pursued, it is necessary in every case to take into consideration not only the size, precise situation, and character of the growth, but also the age, development, and condition of the patient, together with his general and special powers of endurance. If success is not attained by the method first adopted, another may be tried."

Thus it would appear from my table, that the common laryngeal forceps yield a higher percentage of cures than tube-forceps ; but as the latter instrument was used in much more difficult cases, inferences based on the apparent proportionate success would be incorrect.

From the frequency with which a number of instruments have been used in a single case, by other operators,[1] it is much more difficult to give an approximative result of the value of their various methods of treatment ; but the following list will give some idea of the preference for different remedial measures. Thus :—

Forceps were used alone . . . . 13 times.
    „   in combination with other instruments,
        or with caustics, or with both . . 19 „
Knife was used alone . . . . 10 „
    „ with other instruments, or caustics, or
        with both . . . . . . 22 „
Scissors alone . . . . 2 „
    „ with other instruments, or caustics, or
        both . . . . 11 „
Ecraseur or wire loop alone . . . . 34 „
    „ with other instruments, or caustics, or
        both . . . . 14 „
Guillotine, or annular or fenestrated knife
   alone . . . . 8 „
Guillotine, &c., with other instruments, or
   caustics, or both
                    3 „

[1] Appendix D.

Internal treatment may be either mechanical, or chemical, and though in practice it is sometimes necessary to combine these methods, it will be found most convenient, to consider them separately.

MECHANICAL TREATMENT.

Mechanical treatment may be accomplished, (1st) by *evulsion;* (2ndly) by *crushing;* and, (3rdly) by *cutting.* I have not thought it necessary to subdivide the latter process into excision, abscission, and incision, as it would lead to useless repetitions.

Preparatory Measures.—It should be remembered, that, in many cases, before commencing treatment, some previous preparation is required. Congestion of the fauces, elongation of the uvula, enlarged tonsils, and hyperæmia of the larynx, must, if possible, be first subdued by appropriate remedies. Unless the congestion of the larynx be very considerable, it need not be taken into account ; but it is quite useless to attempt any delicate operation on the larynx while the uvula is greatly elongated or the tonsils much enlarged.

Anæsthetics.—In order to facilitate operations on the larynx, various procedures have been recommended for producing anæsthesia of the pharynx and larynx. It is unnecessary, however, to describe the various means recommended, consisting of the application of chloroform, morphia, &c., to the internal parts, the administration of opiates, bromide of potassium, &c., as I have never found any of them of the least use, and some are even dangerous in their effects.[1]

---

[1] Some of the German laryngoscopists recommend repeated pencillings with chloroform and strong solutions of morphia. Their method has been thus described to me :—" At seven o'clock in the evening the larynx of the patient should be painted twelve times with morphia ; at eight o'clock, twelve times with chloroform ; at nine o'clock, twelve times with morphia ; and at ten o'clock, twelve times again with morphia. During the night the patient must be carefully watched, to see *that narcotism is not excessive ;* and, *if necessary, the patient must be stimulated (by strong coffee, 'flipping' with towels, &c.).* At seven o'clock in the morning, twelve more applications of morphia ; at eight o'clock, a laryngoscopic examination is made to ascertain if anæsthesia has been produced ; if sensibility still exist, twelve more applications of morphia must be made ; and so on every hour until the desired condition be established."

Patients cannot, as a rule, be operated on *under chloroform*, unless tracheotomy have been previously performed, or unless the growth be within reach of the finger, or unless it be external to the larynx, viz. in the hyoid fossa or on the posterior surface of the cricoid cartilage. By inhaling a few whiffs of chloroform, however, before treatment is commenced, the larynx is sometimes rendered less sensitive.

By *sucking ice*, also, for a few minutes before the operation, laryngoscopic treatment is more easily borne.

When the epiglottis is long and hangs obliquely, it sometimes hinders operations on the larynx, and several instruments have been invented for raising it. Some continental practitioners even go so far as to pass a thread through the epiglottis, and cause it to be held back by an assistant during the operation. Though such instruments may be useful for purposes of diagnosis, I have not found them applicable where operations have been necessary.

Method of Procedure. — Before introducing instruments into the larynx, they should always be warmed. This precaution should never be omitted, as it greatly diminishes the irritation naturally caused by the use of instruments in the larynx.

As no practitioner would attempt to remove growths without being thoroughly skilled in the use of the laryngoscope and in the application of remedies to the larynx, it is unnecessary to enter into minute details as to the precise mode of carrying out the operation. I may, however, observe, that as, when an assistant holds out the patient's tongue, his hand and arm are apt to get in the way, and the tongue is likely to be drawn to one side, the patient should hold out his tongue himself. In the same way, if it can be avoided, I do not employ an assistant to steady the head ; for this purpose, all that is required is a chair with a high perpendicular back and narrow seat.

(I.) EVULSION is effected with forceps, and is applicable to all growths, except those of cystic character. Cysts have indeed been torn away ; but this is only possible where the

walls are thin and membranous. It is particularly suitable
in cases of sessile growths, for here, other modes of treatment are difficult, and it need scarcely be said that, the softer
the growth, the more favourable it is for this mode of treatment. I am in the habit of removing growths with two kinds
of forceps, viz. the common laryngeal forceps and the tube-
forceps.

Common Laryngeal Forceps.—Common laryngeal forceps
alone have been used by me in 26 per cent. of my cases,
and in conjunction with other measures, in 12 per cent.
more. They are of stout construction, especially as far as
the angle; below this they are somewhat finer, and ter-
minate in expanded, hollow, spoon-shaped, extremities.
The horizontal portion is between 7 and 8 inches in length,
and the part below the angle varies from 2 to $3\frac{1}{2}$ inches.
I formerly employed instruments of much more slender
construction, in order, as far as possible, to avoid loss
of light; but a larger experience has convinced me that
the slight loss of light is more than compensated for,
by the increased steadiness and precision obtainable with
rather stouter instruments. Dr. Fauvel generally employs
forceps with sharp spikes in the blades, and he also has
a catch in the handle, by means of which the two blades can
be locked together. I do not, however, approve of either of
these features, for they both imply the employment of a
greater amount of force than I think desirable. I prefer
rather to desist from evulsion, and resort to excision or
crushing, than to use any considerable degree of force.
Whilst, however, not advising others to use instruments
provided with these additions, I may remark, that I have
frequently seen them employed by my friend Dr. Fauvel
with great success, and with perfect safety.

My common forceps are made in two ways; some open
from side to side (fig. 3, A), others in the antero-posterior
direction [1] (fig. 3, B).

[1] The blades, in my forceps, are connected together by a single screw. I have
not found secondary joints, like those in Cusco's forceps, of any advantage, whilst
they take up more room, and are more likely to irritate the epiglottis.

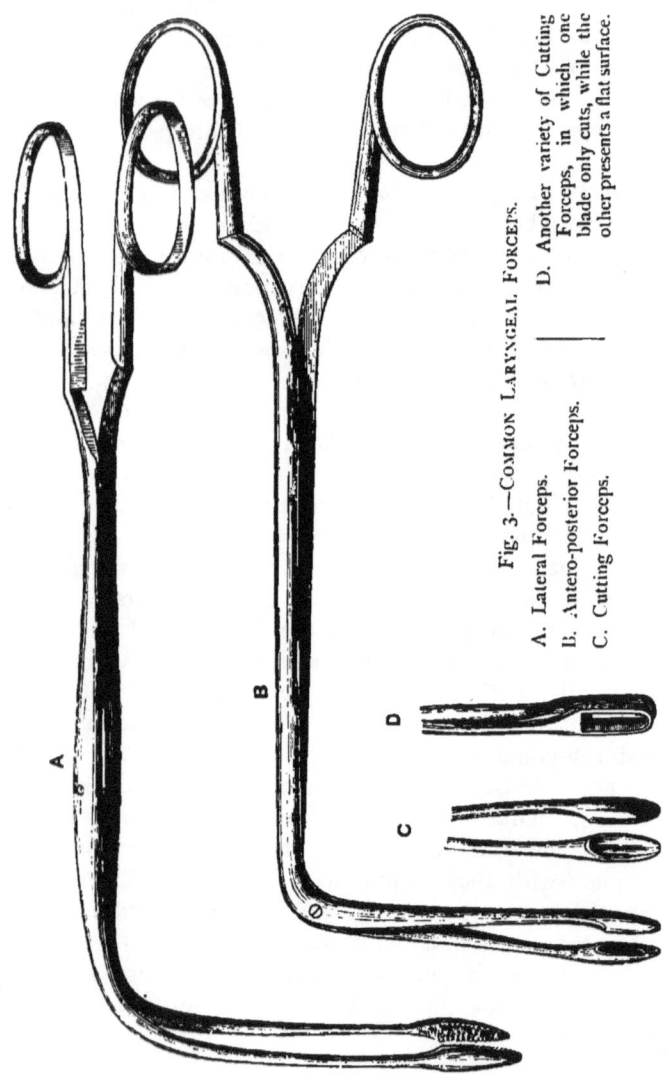

Fig. 3.—Common Laryngeal Forceps.

A. Lateral Forceps.
B. Antero-posterior Forceps.
C. Cutting Forceps.

D. Another variety of Cutting Forceps, in which one blade only cuts, while the other presents a flat surface.

Tube-Forceps.—In the earlier period of my practice I
employed tube-forceps almost exclusively, and it will be
seen that in 34 per cent. of my cases they have been alone
employed. They have been used also in combination with
other remedial means in 18 per cent. of the cases. Mr.
Durham,[1] speaking of these forceps, says, " They are, I
believe, the most generally applicable, and therefore the
best of all instruments yet devised for such purposes."
They are well adapted for a great variety of cases,
especially for those in which the larynx is small, and the
throat irritable.

My tube-forceps (fig. 4) are very slender,[2] and are made
in such a way that they close by the passage of a fine tube
over the shoulders of the blades. The tube is pressed down
by a spring in the upper and anterior part of the handle,
and is easily worked by the index finger when the instrument
is held in the hand. The instrument can be made shorter
or longer at will, and the blades can be made to open
either laterally or in the antero-posterior direction. Various
blades (fig. 4, 1, 2, 4) and scissors (fig. 4, 3) can be fixed
to the same stock, according to the requirements of the
operator.

Tube-forceps were introduced into Germany soon after the
invention of the laryngoscope, and they were first recommended by Semeleder, at about the same time, or soon after, I
commenced to use them in England. At a later period Stoerk
employed a somewhat similar instrument (fig. 6, D, E, F).
In the German instruments, however, the blades of the forceps
are made to close by being partially drawn within the tube,
whilst, as already described, in my instrument the blades of
the forceps are the fixed point, and the tube passes over
their shoulders. The obvious advantage of my instrument
will be at once evident when it is remembered that the
operator has to seize a small body, seen only by reflected
light. In my instrument, also, the arrangement by which

[1] *Op. cit.*, p. 585.
[2] From the manner in which the instrument acts, delicacy of construction does
not cause vibration or unsteadiness.

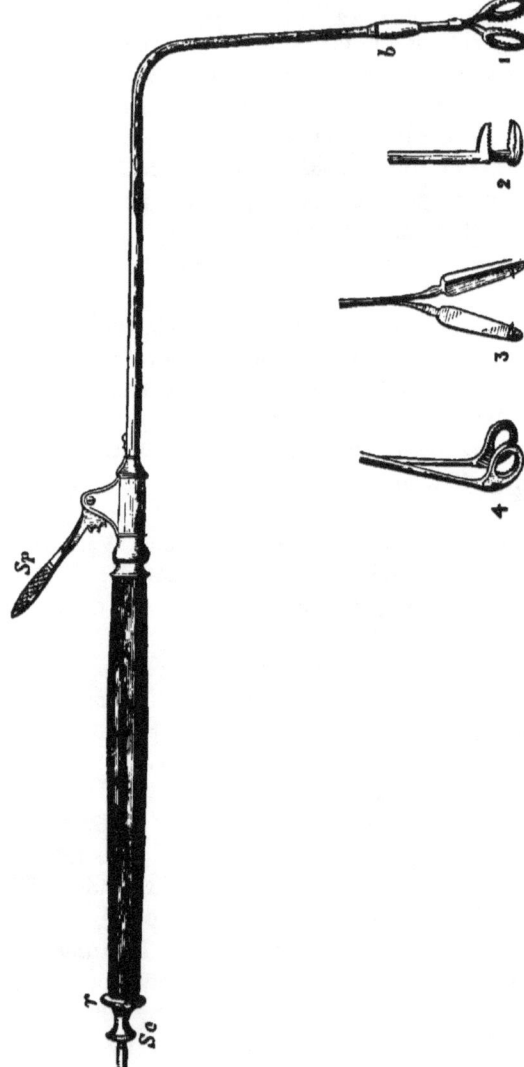

Fig. 4.—The Laryngeal Tube-Forceps, and Scissors.

Sp. The Spring, by pressing on which, the tube is forced over the base of the forceps.

b. The Joint at which longer or shorter tubes may be applied, and the blades taken out and cleaned (this part has been made unnecessarily thick by the engraver).

r. The Ring, by turning which the forceps revolve, so that the blades open in any direction.

Sc. Screw for taking the Instrument to pieces, cleaning it, &c.

1. The Perpendicular Blades.

2 and 4. The Horizontal Blades.

3. The Scissors, with hooks attached to them.

the closure is effected is more conveniently placed in the handle of the instrument, so that greater steadiness is secured, and the hand of the operator does not obstruct the light.

(II.) CRUSHING can be carried out with either of the two kinds of forceps already described, and has been used, in conjunction with other methods, in 3 of my 100 cases: it has also been employed by Lindwurm, Schroetter, Türck, and others, and has been used in 5 per cent. of the cases treated by other practitioners (Appendix D). I formerly employed this plan of treatment in cases in which the growth was of dense structure, and very firmly attached; but latterly I have generally used cutting instruments in these cases. Crushing, however, is preferable to using force in evulsion. As a rule, the stronger kind of forceps are required; but the blades should be flatter, *i.e.* less spoon-shaped, and rougher, than for evulsion. The American translator of Dr. Tobold's work[1] describes the process as "crushing up," and observes, that energetic and repeated compression of the tissue is all that is required, to destroy the conditions of nutrition and produce mortification, and that subsequently the dead portion can be separated. It is probable, that, in many cases, where evulsion is adopted, crushing, at the same time, takes place; in other words, that, when a growth is torn away, its base is, to a greater or less extent, lacerated and crushed. The success of evulsion must, therefore, in part, be attributed to the incidental crushing which takes place.

(III.) CUTTING may be carried out, as already remarked, either by excision, abscission, or incision. For incision, cutting forceps are used; abscission may be performed by means of knives, scissors, guillotines, or écraseurs; while for incision, or scarification, knives or lancets are employed. The various cutting instruments will now be described.

---

[1] *Chronic Diseases of the Larynx*, by George M. Beard, A.M., M.D. New York, 1868. This translation has been faithfully made by the accomplished neurologist of New York, and can be thoroughly recommended.

Cutting Forceps.—Though in my 100 cases cutting forceps have been employed in only two instances, I have of late used an instrument of this sort very frequently and with great success, and I believe that it will be very serviceable in future. Dr. Elsberg has also employed similar instruments for some years, and they have lately been employed by Dr. Schroetter. One form of my cutting instrument consists of the ordinary forceps, but each blade is deeply spoon-shaped, and the margins of the spoons have very sharp edges (fig. 3, C). In another, the cutting edge of one blade closes against a flat disc of lead, which is contained in the opposite blade (fig. 3, D). These instruments are most useful in cases in which the growth is very hard. In illustration of their value, I may refer to a case [1] in which the tumour could be easily seized with the common forceps, but, owing to its extreme density, only very small portions could be removed ; crushing likewise proved unavailing, and attempts at removal caused such severe dyspnœa that tracheotomy became necessary ; subsequently, however, on using the cutting forceps, I was able to remove the growth with ease.

Knives.—I have employed the knife in only three per cent. of my 100 cases (not including two cystic cases); but they have been much more extensively used on the continent. Thus excluding English cases from Appendix D, I find that knives were used in more than 25 per cent. The knives (fig. 5) used may be either guarded or free. In operating on cystic growths, where a comparatively large opening has to be made, a cutting *blade* is required ; but for incising small fibromata, growing from the edge of the vocal cords, it is often more easy to puncture or stab the growth at its point of insertion, and for the latter purpose the *trocar-pointed knife* of Tobold is very convenient. I formerly employed the guarded lancet, but now almost invariably use the naked instrument.

Scissors or Shears.—These instruments are of very little use for the removal of laryngeal growths. Though employed

---

[1] Case 74, Appendix A.

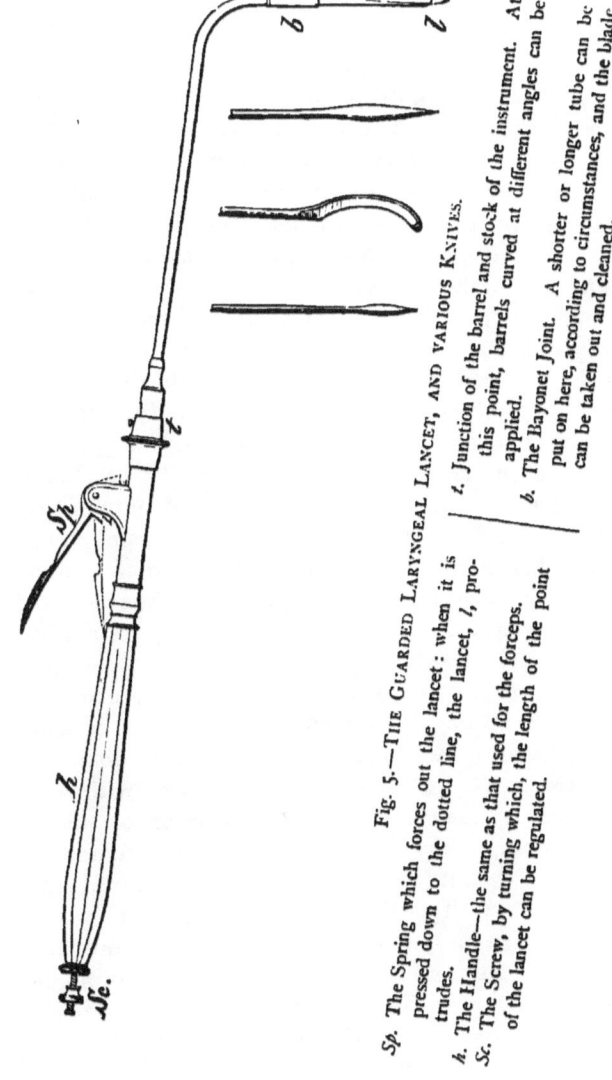

Fig. 5.—THE GUARDED LARYNGEAL LANCET, AND VARIOUS KNIVES.

Sp. The Spring which forces out the lancet: when it is pressed down to the dotted line, the lancet, *l*, protrudes.

h. The Handle—the same as that used for the forceps.

Sc. The Screw, by turning which, the length of the point of the lancet can be regulated.

t. Junction of the barrel and stock of the instrument. At this point, barrels curved at different angles can be applied.

b. The Bayonet Joint. A shorter or longer tube can be put on here, according to circumstances, and the blade can be taken out and cleaned.

when laryngoscopy was in its infancy, they have now almost fallen into disuse. Scissor-blades (fig. 4, 3) can be attached to my tube-forceps, but I have never been able to employ them with success; for in order that scissors should act efficiently, it is necessary, first, that the substance to be cut, should have a certain degree of firmness, or tension, and secondly, that the blades should bite well in relation to each other. Neither of these conditions exists; for whilst the consistence and mobility of the growth render it difficult to be cut, the length, delicacy, and angular form of the instrument, as well as the mode by which the blades are, of necessity, made to approximate, are all adverse to its action. In addition, the blades of scissors cannot, of course, be made to open so widely as those of forceps, and it is necessarily much more difficult to seize a growth with a thin-edged instrument like scissors, than with the broad spoon-shaped extremities of forceps. Occasionally, where it has not been possible to entirely cut through a growth, incisions have been made into its structure: like crushing, this procedure has the effect of disturbing the nutrition of the neoplasm, and thus favours its degeneration by sloughing or atrophy. This treatment is only mentioned in consequence of its having been employed with success in Professor Bruns' first case: it is now entirely superseded by other and more certain methods.

Ecraseurs.—It may, perhaps, be thought that the écraseur does not act as a cutting instrument; but, in point of fact, the wire employed is so fine, that it actually does *cut* through the tissue. Growths have been removed with considerable success by means of the wire écraseur or noose. It was first employed by Dr. Walker,[1] of Peterborough, who was also the first practitioner, in this country, to extirpate a growth with the aid of the laryngoscope. His instrument was made on the principle of Gooch's canula, but bent at right angles, and much longer below the angle. Subsequently Sir Duncan Gibb invented a more convenient écraseur, in which the wire is drawn through eyes at the

---

[1] *Lancet*, November, 1861.

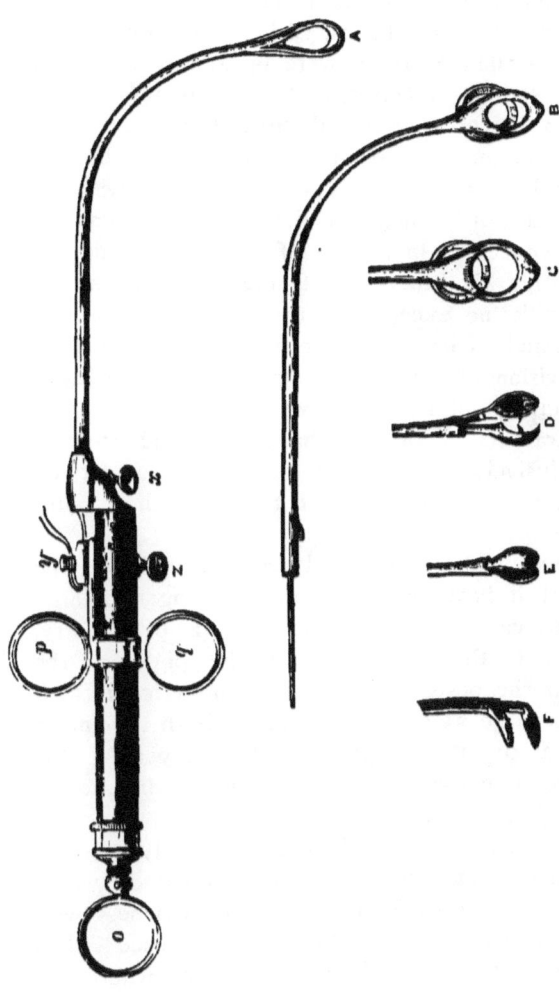

Fig. 6.—Stoerk's Instruments.

A. Ecraseur.

B, C. Guillotines of various sizes.

D, E, F. Forcep-blades of different kinds.

In all cases, the instrument acts through an internal rod, a double wire being drawn through a tube-shank, which is attached to the handle at *x.* The movable part of the instrument is attached by the screw, *z*; and, in the case of the écraseur, the ends of the wire are twisted round the peg, *y*. In using the instrument, the operator puts his thumb into the ring, *o,* and his index and second fingers through the rings, *p* and *q,* and when traction is made on these rings, the écraseur or forceps are drawn upwards.

extremity of a curved metallic rod, instead of through the two canulæ of Gooch's instrument. This écraseur, at a later period, was still further improved by Dr. George Johnson. Treatment by écraseur has also occasionally been carried out by Trélat, Moura-Borouillou, Bruns, Elsberg, and others.

The various so-called écraseurs were, up to a certain time, all open to the serious objection, that the wire was unprotected, and, being necessarily very flexible, it was very often pushed or bent on one side before the loop could be put round the growth. This difficulty was subsequently overcome by Dr. Stoerk, of Vienna, who first suggested that the wire should be concealed in a loop of rigid metal (fig. 6). It will be seen that Dr. Stoerk's instrument has almost the character of a guillotine, except that a wire is used instead of a cutting blade, and, as has already been shown, in all these instruments, division of the growth is accompanied by a quick cutting action, not by the slow squeezing movement of the true écraseur. Though I have not used Stoerk's instrument exclusively in any one case, I have derived assistance from it in 4 per cent. of the cases here published.[1]

I have, however, had a true wheel écraseur made for the larynx, and with it have succeeded in removing two large growths. In one case,[2] the growth was the size of a cherry, and was attached to the under surface of the epiglottis; the other grew from the posterior surface of the cricoid cartilage, and was as large as a bantam's egg. Both specimens were exhibited at the Pathological Society last year.[3] One case is reported hereafter,[4] but the other proving to be epithelioma, has been excluded.

My "guarded wheel écraseur" (fig. 7) is only adapted for large growths, and is especially indicated where hæmorrhage may be anticipated. From the slowness with which the instrument acts, it can only be used when tracheotomy has been previously performed, or where the

---

[1] Appendix A, Cases 57, 79, 84, and 86. [2] Appendix A, Case 88.
[3] *Pathological Transactions*, vol. xxi. p. 51, *et seq.*
[4] Appendix A, Case 88.

growth is external to the laryngeal canal (*i.e.* on the posterior wall of the cricoid cartilage or in the hyoid fossa, &c.). In both my cases, the operation was done under chloroform.

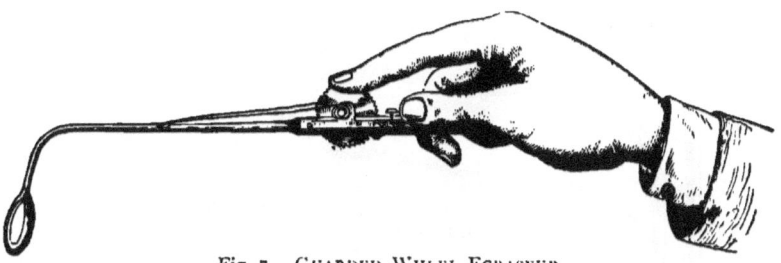

Fig. 7.—GUARDED WHEEL ECRASEUR.

Loops and Rings.—Loops or rings of rigid wire (fig. 8) can sometimes be employed with success. They should be curved at different angles, and be of different lengths. The opening of the loop should sometimes be in the antero-posterior, sometimes in the lateral direction, and sometimes intermediate between these two, so that different loops can be employed, according to the situation of the growth. It is convenient to let the inner edge of the wire be sharp, like a knife, and a kind of angular collar is not without its advantages. By means of these instruments a growth may occasionally be jerked, or scraped, off.

No Danger in Ecraseurs or Wire Loops.—In the use of the small ordinary cutting écraseurs I formerly feared that portions of the growth, separated in the operation, might fall down the windpipe, and give rise to serious irritation of the lungs; but experience has convinced me that this danger is chimerical, and that these instruments, if perhaps less useful than others, are nevertheless perfectly harmless.

Guillotines.—Incision by guillotine was recommended at an early epoch of laryngology by Drs. Ozanam, Semeleder, Stoerk, and others ; but it is not at all a convenient mode of treatment. Not only is it difficult to press the instrument sufficiently against the parietes, but with a long slender curved

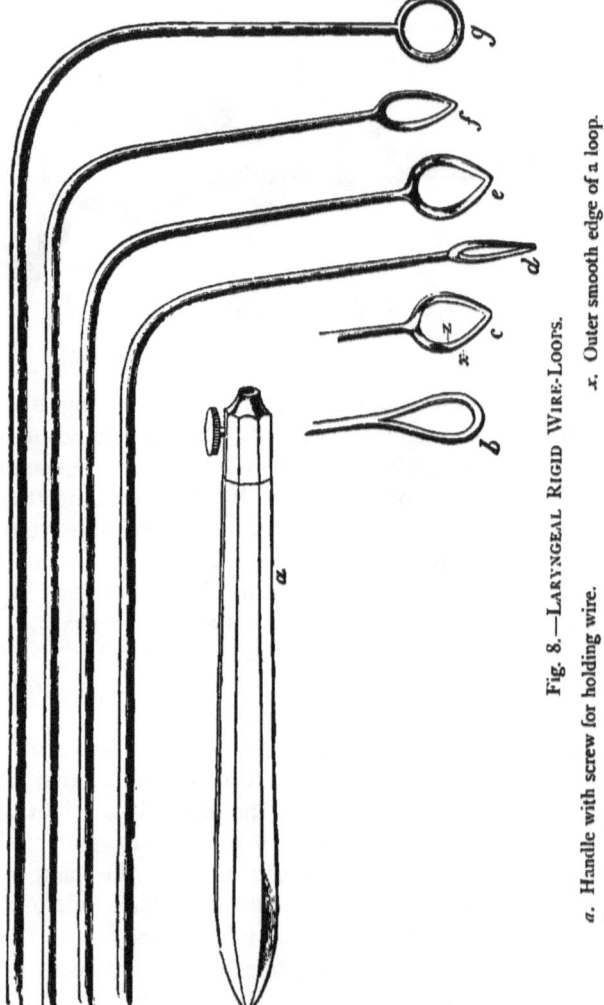

Fig. 8.—Laryngeal Rigid Wire-Loops.

a. Handle with screw for holding wire.
b, c, d, e, f, g. Different wires showing different kinds of loops.

x. Outer smooth edge of a loop.
z. Inner cutting edge of a loop.

instrument it is impossible to get the sudden forcible impulse which is required for a guillotine. The late Dr. Türck invented what he called a fenestrated knife (fenstermesser). This differed from Semeleder's guillotine, inasmuch as the protecting sheath of the blade was made square, instead of round, and the cutting blade, instead of being straight, was lancet-shaped, so that it presented two cutting edges. It does not appear to me to have been in any respect superior to the ordinary guillotine, and neither instrument is well adapted for the removal of laryngeal growths. Only very small guillotines can be used, and only a very small portion of a growth can, as a rule, be sliced off. I have never been able to employ these instruments with advantage.

CHEMICAL TREATMENT.

Chemical treatment may be carried out either with caustics, escharotics, or galvanic cautery.

Caustics.—Solutions of nitrate of silver are generally of but little use ; if employed, however, they should be exceedingly concentrated, and should be accurately applied, with a very fine camel's hair pencil, to the seat of disease. On reference to my own cases,[1] but especially to those treated by other practitioners,[2] it will be seen that when laryngoscopy was first introduced, growths were generally treated by the application of caustics. This was no doubt due to the circumstance, that at that period, practitioners were not aware, to how great an extent operations could be conducted within the larynx, and at that time, of course, no great manual dexterity in this department had been acquired. The small utility of this treatment is, however, demonstrated by the fact that since 1862 mechanical methods have almost entirely superseded the local application of caustics. In some of the earlier cases, the apparent success of caustic treatment was also probably due to defective

---

[1] Appendices A and C.   [2] Appendix D.

diagnosis, cases of condylomata and inflammatory thickening having been mistaken for true growths. I, myself, have only seen caustic solutions of use in three cases.[1] In one instance the growth was entirely destroyed, but in the others some diminution alone took place. On the other hand, I know of numerous cases, in which solutions of nitrate of silver have been applied persistently for several months, without producing any effect on the growth. The solid nitrate of silver is equally unavailing; but it may sometimes be applied with advantage to the remains of a growth which has previously been nearly removed by evulsion. In treating cystic growths, it is a good plan to apply caustic to the interior, after an incision has been made, and the contents of the cyst evacuated. The most convenient mode of applying it, is, to fuse a small quantity of the salt on to a piece of aluminium wire of suitable length and curve.

Chromic Acid in my hands has proved as unsatisfactory for the destroyal of growths, as solutions of nitrate of silver.

Escharotics.—On a few occasions I have employed escharotics with marked success, but only in a supplementary way. They may be used in cases, where numerous small growths cover a large surface of the mucous membrane of the larynx.[2] I have occasionally employed nitric acid, but the escharotic which I have found most useful is " London paste."[3] To all caustics and escharotics, however, the objection remains, that if sufficiently powerful to be effective, they are very likely to cause spasm of the glottis, or to give rise to inflammation of the adjacent mucous membrane; for this reason I now very seldom use them.

Galvanic Cautery.—Galvanic cautery may be carried out,

---

[1] Appendices A and C, Cases 1, 2, and 54.

[2] Appendix A, Case 3.

[3] This preparation consists of equal parts of caustic soda and unslaked lime. It should be kept as a powder, and only made into a paste with water when required for use. It differs from Vienna paste, in so far as soda is employed instead of potash, and water is used for mixing in place of alcohol : it is a far more manageable and less painful compound. It should be applied on the point of an aluminium rod, as I have recommended for nitrate of silver.

either with knife-like instruments, or with loops. This plan of treatment was first practised by Professor Middeldorpf, and has since been very successfully carried out by Dr. Voltolini, of Breslau, and other practitioners. Notwithstanding, however, that I have modified the latter physician's laryngeal electric-cautery instruments, and have had them made exactly on the principle of my laryngeal electrode, I cannot say that I have found the treatment well adapted for the destruction of growths. Professor Bruns strongly insists on the disadvantages of galvanic-cautery treatment, and notices how "much more difficult it is to limit the caustic action of the glowing platinum loop to a small surface, than it is, when nitrate of silver is used." I have only employed galvanic cautery exclusively in three cases, and once in conjunction with other means.[1] In one instance the patient complained very much of the painful character of the treatment, and in two cases it even gave rise to acute œdema, though fortunately of a very localized character. In addition to the inconvenience it causes to the patient, galvanic cautery gives a great deal of trouble to the practitioner. As the wires have to be very carefully isolated, the electrode is of course rather bulky, and the necessarily unwieldy covered wires, which are attached to the handle of the electrode, prevent the operator using the instrument with that delicate precision which is essential in all manipulations on the larynx. When to this is added the employment of strong acids and a special electric apparatus, which requires some experience in its use, and the presence of an assistant at each operation, it will be seen that the employment of electric cautery introduces a number of complications and difficulties in the way of the operator. Further, as an instantaneous white heat ought to be produced, the wire at the extremity of the instrument must be exceedingly fine, and is therefore constantly destroyed by combustion. Hence it happens that the instrument is continually in the hands of the instrument-maker. There are,

---

[1] Appendix A, Cases 38, 41, 42, and 49.

indeed, no special advantages to be derived from this
method : the other modes of treatment are amply sufficient,
and there is no particular kind of growth for the destruction
of which electric cautery is indicated.   It will be seen, therefore, that, as far as my experience goes, this treatment is
not to be recommended.

Extra-Laryngeal Methods of Removing Growths.

In certain cases, it unfortunately happens that growths in
the larynx cannot be removed through the mouth.   The
introduction of the laryngoscope, by rendering the diagnosis
of laryngeal growths for the first time possible, and even
easy, has not only led to the valuable laryngoscopic treatment already described, but it has given an increased
prospect of success to the various surgical operations by
which the cavity of the larynx is laid bare.   Not only
does the laryngoscope reveal the situation of the growth,
but its extent and nature can generally be ascertained.
These circumstances enable the surgeon to adapt the
operation to the special circumstances of the case, and
teach him to avoid operating in those of an unsuitable
character.

The difficulty of laryngoscopic treatment may be due to
the large size or extreme density of a growth, to its inacces-
sible situation, or extensive origin ; to the occurrence of
inflammatory tumefaction, or spasm of the glottis, on attempted evulsion through the mouth ; to great irritability of
the fauces, or to an unusually nervous and excitable state of
the patient.   In the case of very young children also, an
extra-laryngeal method may be necessary.

The large size of a growth does not, in itself, call for
external treatment, some of the largest growths having
been removed *per vias naturales.*[1]   The extreme density of a

---

[1] Appendices A and C, Cases 3, 52, 92, 95, &c.

growth sometimes presents a great difficulty to laryngoscopic treatment, but with strong cutting forceps, this difficulty is only insuperable in the case of ecchondroses, and it is very questionable whether radical treatment should be attempted for their removal. The growth may be so situated that it cannot be completely eradicated from above. This occasionally happens in the case of growths springing from the anterior wall of the larynx below the vocal cords. In one of my cases of this sort, the evulsion was incomplete,[1] but in two others the growth was entirely eradicated. When a growth, however, is situated in the ventricle, and only slightly projects from the ventricular orifice, it is sometimes impossible to remove it entirely from above. The projecting portion may be cut off, but the base remains. The question arises in these cases, whether the growth should be removed from time to time, as it sprouts out of the ventricle, or whether an external operation should be performed. Should a large tumour gradually extend the ventricle and push the ventricular band before it, direct incision into the larynx, and subsequent incision into the ventricle, might possibly be called for; but as a matter of fact, as soon as the ventricle becomes at all tense, the growth emerges from the ventricular orifice. Indeed, even before the ventricle becomes at all full, the disposition to growth naturally takes place at that point where the resistance is least, and neoplasms, originating in the ventricle, emerge at an early period from its cavity.

The extensive origin of growths greatly increases the difficulty of their removal, especially when there is an independent multiple development. Though this occurred in 3 of my cases, I was able to remove the growths through the mouth in 2 of them.[2] Difficult as it is, in such cases, to effect evulsion *per vias naturales*, it would probably be not less difficult to accomplish complete eradication by an external operation.

---

[1] Appendices A and C, Case 24.
[2] Appendices A and C, Cases 3, 46, and 98.

The occurrence of inflammation or spasm of the glottis, on attempted laryngoscopic treatment, may render the *combined method* necessary (tracheotomy having first been performed, and evulsion being subsequently effected through the fauces), but it does not in itself justify an extra-laryngeal operation for evulsion.

An insuperable irritability of the fauces, or an extremely nervous condition of the patient, may, however, render laryngoscopic treatment impossible ; and in these cases an extra-laryngeal treatment may be necessary. In the case of young children who cannot be taught to submit to laryngoscopic treatment, extra-laryngeal treatment may be required ; but it must not be forgotten that very young children have been successfully treated with the aid of the laryngeal mirror. In one of my cases,[1] the patient was only four years old, and in another [2] only six. In a case [3] treated by Professor Bruns, the child was only five years old, and though the neoplasms showed an extreme disposition to exuberant growth and multiplication, portions of the neoplasms were removed through the mouth and others destroyed with caustics.

Contra-Indications for extra-Laryngeal Methods.—It may be stated as a cardinal law, that *an extra-laryngeal method ought never to be adopted* (even where laryngoscopic treatment cannot be pursued) *unless there be danger to life from suffocation or dysphagia.* Direct incision into any part of the air-passages is always attended with both immediate and remote danger to life, the amount of risk being dependent on the situation of the opening, and on the mode in which the treatment is carried out. The various degrees of danger will be considered when discussing the merits of the different extra-laryngeal methods, and it is sufficient here to remark, that the existence of dysphonia does not justify operations,

---

[1] Appendix A, Case 11.   [2] Appendix A, Case 12.

[3] Appendix D, Case 89. In this case, tracheotomy had been performed some years before the little patient was placed under Dr. Bruns' care, but his treatment was mainly carried out through the mouth, with the aid of the laryngeal mirror.

which, though easy to perform, may be regarded as " capital." Hence an extra-laryngeal operation is not justifiable for the removal of a *small* growth in the larynx, unless that growth give rise to dangerous dyspnœa, and cannot be removed by a less serious method. No doubt a direct incision into the larynx offers a readier and more brilliant method of extirpation than the more tedious process of laryngoscopy, but to do an external operation where laryngoscopic treatment could be ultimately successful is not less reprehensible than to perform lithotomy in a case of calculus, in which lithotrity could be effected, or to amputate a limb where resection could be accomplished. A dangerous operation, even though successful, is not justifiable, when a perfectly safe method might have been adopted. At least one case, if not more, has been reported in which the growths might easily have been removed through the fauces, and in which neither the size of the growth nor its situation justified the external operation.[1]

If these improper cases were withdrawn from the various statistics in which they appear, the results of external incision would appear much more unfavourable. Contra-indications based on danger to life, having been thus briefly pointed out, it only remains for me to remark that destruction of the vocal function is often the result of the most available extralaryngeal method.

Extra-laryngeal methods of extirpation may be carried out in one of three ways :—1st, By division of the thyroid cartilage, or thyrotomy ; 2ndly, by supra-thyroid laryngotomy, or division of the thyro-hyoid membrane ; and 3rdly, by infra-thyroid laryngotomy (through the cricothyroid membrane), or tracheotomy.

---

[1] It is certainly remarkable that there should have been seven cases published from the small town of Buda-Pesth with its 200,000 inhabitants ; whilst from London and Paris, with their 5,000,000 inhabitants, only six cases have been as yet recorded.

Division of the Thyroid Cartilage, or Thyrotomy.

Indications for Operation.—This operation may be required for the removal of large growths in the cavity of the larynx, causing great dyspnœa or dysphagia, which cannot be removed with the aid of the laryngoscope, or for the evulsion of growths in the sub-glottic region, which cannot be extirpated by indirect laryngotomy (through the crico-thyroid membrane). It might be thought that this operation would be called for in the case of children; but the facility with which they can be treated laryngoscopically has already been pointed out; and it must not be forgotten that when the larynx is small, thyrotomy is much more likely to lead to injury of the vocal cords. It may be remarked that in several of the cases, hereafter reported, the disease having been cancer, the operation was of very doubtful utility; whilst in others, the growths were small, and with patience and perseverance could probably have been removed through the mouth. Ossification of the cartilage is in no sense a contra-indication for the operation. It adds little to its difficulty, though union afterwards takes place rather more slowly.

History.—This important operation was first proposed for the removal of laryngeal growths by Desault, at the end of the eighteenth century. His remarks, which were perfectly true before the invention of the laryngoscope, are as follows :—

"In cases of polypi of the larynx, the indications are twofold; viz. the extirpation, or ligature of the growth, and the re-establishment of a passage for air; and they both necessitate laryngotomy. It rarely happens, indeed, that laryngeal excrescences project so far into the mouth, that they can be seized and extirpated or ligatured *per vias naturales.*"[1]

---

[1] This quotation is taken from a later edition of Desault's *Œuvres chirurgicales*, by Bichat. Paris, 1812, vol. ii. p. 255.

The operation was not, however, carried out till the year 1833, when it was performed for the first time by Brauers of Louvain. Ten years later it was repeated by Ehrmann of Strasbourg. In 1851 it was practised by Gurdon Buck, and again by him in the year 1861. The invention of the laryngoscope naturally gave an impetus to this operation, and its progress will be at once evident on reference to the Thyrotomy Table on pages 92 and 93.

Method of Procedure.—The first question which arises is whether tracheotomy should or should not be performed as a preliminary measure of safety. In some cases, this precaution has been taken, whilst in others it has not been adopted. Tracheotomy should, in my opinion, always be performed, in the first instance. Three advantages are thereby gained, and no additional danger is incurred :—1st, the danger of suffocation, from blood passing into the trachea, is avoided ; 2ndly, the operation can be performed with greater deliberation, and the growth can be removed with more certainty ; 3rdly, if laryngitis should subsequently supervene, the patient is in a condition of safety.

The danger of suffocation during the operation is by no means chimerical ; for although, in some cases, no evil has resulted from the neglect of this precaution, in two instances [1] in which the operation was attempted without tracheotomy, the patients almost died from blood passing down the windpipe, and in both cases, an incision had to be quickly made from the second ring of the trachea to the thyro-hyoid membrane. If a small growth the size of a pea or tare be removed by thyrotomy, it may not be so necessary to perform tracheotomy in the first instance ; but as I do not consider that an extra-laryngeal method is justifiable for the extirpation of small growths, such cases are excluded from consideration. When the patient has recovered from tracheotomy, that is to say when all bleeding, coughing, and spasmodic movement of the larynx have ceased, the incision should be made exactly in the median line, through the

---

[1] Thyrotomy Table, Nos. 19 and 23.

textures over the thyroid cartilage, from the thyroid notch to the upper border of the cricoid cartilage. The thyroid cartilage should then be most carefully divided by a succession of small nicks, with a short, strong, sharp-pointed knife; but if ossification have taken place, the opening must be effected with a small circular or convex saw. The instrument should not be allowed to penetrate the larynx until the whole of the cartilage is divided.[1] By this method the paroxysms of coughing, which otherwise interfere with the operation, are often avoided. When divided, the alæ of the cartilage should be kept widely apart by means of strong retractors held by two assistants, one on each side of the patient. The retractors[2] should be like miniature pitchforks, with the points blunted and bent round, so that they can hold back the alæ.

If the alæ cannot be thrown back, the crico-thyroid membrane should be divided along the lower edge of the thyroid cartilage, on one side, or, if necessary, on both sides. If there be still insufficent room, the thyro-hyoid membrane should be divided, by a horizontal section along the upper edge of the thyroid cartilage. Horizontal division of the membranes, however, is not generally necessary, and the thyro-hyoid should if possible be left intact.

The operator should now throw a strong reflected light into the opening, and, guided by it, and his previous laryngoscopic knowledge of the case, he will be able to seize the growth with a hook or forceps, and divide it with a pair of short curved scissors. On account of the small space at the command of the operator, the growth may sometimes be divided, without being previously seized, or it may be torn away with forceps. The base of the growth should then be firmly touched with solid nitrate of silver. Actual cautery, acid nitrate of mercury, and galvanic cautery, have all been used, but I prefer the nitrate of silver, as less likely to give rise

---

[1] This precaution is justly insisted on by Krishaber and Planchon (*Faits cliniques de Laryngotomie*. Paris, 1869, p. 93).

[2] This form of retractor is better than a solid plate of metal, as, when in use, it does not hide the laryngeal surface.

to laryngitis, and quite as effectual when applied to a raw surface.

The two alæ of the thyroid cartilage should then be carefully brought together, in their exact normal situation, with two silver sutures, and the edges of the wound united with plaster. The canula should be allowed to remain in the trachea, for, at least, a few days, until all danger has passed off; or if there be any likelihood of recurrence, till further steps have been taken to effect complete eradication.

In some of the cases contained in the annexed Table, the cricoid cartilage was divided, and though no harm appears to have resulted from its section, it is better, if possible, to leave it intact. This precaution is, I believe, one of considerable importance in relation to the subsequent result of the operation; for as it is often necessary that the patient should wear the tracheal canula at least for a few days after the operation, and sometimes for weeks or months, if the cricoid cartilage has been divided, the tube is likely to be constantly pushed up against the thyroid cartilage in coughing, and this pressure is almost certain to interfere with the proper union of the two parts of the thyroid cartilage. Krishaber[1] justly remarks that division of the cricoid cartilage is altogether *unnecessary;* for whilst, on the one hand, it does not facilitate the removal of growths above the vocal cords, those below the glottis can easily be removed either through an opening in the crico-thyroid membrane or the trachea.

Comparative Merits of Thyrotomy.—Unlike the operation conducted *per vias naturales*, in being unattended with danger to life, or risk of destruction of function, the procedure now under consideration is a very serious one. Its results will be best appreciated by an investigation of the annexed Table.

[1] *Op. cit.*

THYROTOMY TABLE,

SHOWING RESULTS, AS REGARDS LIFE AND VOICE, IN ALL CASES OF LARYNGEAL GROWTH, WHETHER MALIGNANT OR BENIGN, IN WHICH THYROTOMY WAS PERFORMED.[1]

No	Date.	Operator.	Operation.	Result.
1	1833	Brauers, Louvain ... (Ehrmann, *Op. cit.*, p. 12.)	Thyrotomy without Tracheotomy.	Death.
2	1844	Ehrmann, Strasbourg (*Op. cit.*, p. 18.)	Thyrotomy with Tracheotomy (Cricoid divided).	Persistent Aphonia. (Death from typhus 7 months later.)
3*	1851	Gurdon Buck, America... ... ... (*New York Medical Journal*, May, 1865)	Thyrotomy with Tracheotomy and partial removal of growth (Cricoid divided).	Death 18 months later. Persistent Aphonia. Canula always worn.
4	1861	Rauchfuss, St. Petersburg.	Thyrotomy with Tracheotomy (Cricoid divided).	Death in 2 years. Persistent Aphonia.
5	1861	Gurdon Buck, America...	Thyrotomy, preceded by Tracheotomy 5 months previously (Cricoid divided).	Persistent Aphonia. Recurrence of growth. Canula always worn.
6*	1862	Sands, America ... (*New York Medical Journal*, May, 1865.)	Thyrotomy with Tracheotomy (Cricoid divided).	Death 22 months later, from malignant disease of supra-renal capsules. Voice never normal.
7	1863	Boeckel, Strasbourg...	Thyrotomy with Tracheotomy (Cricoid divided).	Death in 2 months, from Pneumonia. Persistent Aphonia.
8	1863	Busch, Bonn	Thyrotomy with Tracheotomy (Cricoid divided).	Persistent Dysphonia. Recurrence of growth. Canula still worn.
9	1863	Debrou, Paris	Thyrotomy with Tracheotomy (Cricoid left intact).	Death in 7 days, with metastatic abscesses in both lungs.
10*	1864	Gibb and Holthouse, London. (*Brit. Med. Journal*, Sept. 30, 1865.)	Thyrotomy, preceded by Tracheotomy 8 days previously (Cricoid divided).	Death in 1 year. Persistent Dysphonia. Recurrence of growth.
11	1864	Lewin & Ulrich, Berlin.	Thyrotomy with Laryngotomy.	Cure. "Voice much deeper after operation."
12	1864	Gilewsky, Cracow ...	Thyrotomy only (Cricoid left intact).	Persistent Dysphonia.
13	1864	Gouley, America ...	Thyrotomy, preceded by Tracheotomy 10 weeks previously (Cricoid divided).	Persistent Aphonia. Normal respiration.
14	1865	Balassa, Pesth	Thyrotomy with Tracheotomy (Cricoid divided).	Cure.
15	1865	Koerberlé, Strasbourg	Thyrotomy only	Persistent Aphonia. Still wears canula.
16	1866	Balassa, Pesth	Thyrotomy only	Cure.
17	1866	Ditto	Thyrotomy with Tracheotomy (Cricoid divided).	Cure, though growth was incompletely removed.

No.	Date.	Operator.	Operation.	Result.
18	1867	Balassa, Pesth	Thyrotomy with Tracheotomy (Cricoid divided).	Great benefit, but removal of growth incomplete. Thyrotomy performed a second time without Tracheotomy. Ultimate Cure.
19	1867	Cutter,[b] America ...	Ditto ditto	Persistent Dysphonia. Recurrence of growth in less than a month after operation.
20	1868	Mackenzie and Couper, London.	Ditto (Cricoid left intact).	Cure. Recurrence 2 years later.
21	1868	Mackenzie and Evans, London.	Ditto ditto	Persistent Aphonia, but respiration normal.
22	1868	Navratil,[c] Berlin ...	Thyrotomy only (Cricoid left intact).	Persistent Dysphonia.
23	1868	Ditto[d]	Thyrotomy with Tracheotomy (Cricoid divided).	Growth not removed.
24	1868	Ditto[e]	Thyrotomy only (Cricoid divided).	Cure.
25*	1869	Schroetter, Vienna ... (*Medicin Jahrbücher*, Wien, 1869, vol. xvii. 2nd Heft, p. 81.)	Thyrotomy with Tracheotomy (Cricoid divided).	Death from hæmorrhage in 7 hours.
26*	1869	Mackenzie and Wordsworth, London ...	Thyrotomy, preceded by Tracheotomy 14 days previously (Cricoid left intact).	Death in 6 months. Persistent Dysphonia.
27*	1869	Cohen,[f] America ... (*New York Medical Record*, Aug. 16, 1869.)	Thyrotomy with Tracheotomy.	Condition of voice not stated. Recurrence began to manifest itself a fortnight after operation.
28	1862	Krishaber, Paris ...	Thyrotomy with Tracheotomy (Cricoid left intact).	Cure.

[1] The reference applicable to each case, where it is not inserted in the text, will be found in Appendices A or D.
* These six cases were malignant or semi-malignant.
[a] In May, 1851, section of thyroid and cricoid cartilages and trachea, removal of growth and insertion of canula. In the following September, section up to the hyoid bone and an inch below permanent orifice in trachea. Second removal of growths, re-insertion of canula. In January, 1852, tracheotomy performed lower down. In August patient was accidentally suffocated in changing canula.
[b] Thyrotomy was performed in the first instance; but as the patient almost died under the operation, from blood passing into the trachea, an extensive incision had to be made from the upper rings of the trachea to the thyro-hyoid membrane. Improvement in voice is reported; but as the growth recurred in less than a month, persistent *aphonia* would probably more correctly describe the condition.
[c] The right vocal cord was wounded in the operation, and a portion of it afterwards sloughed away. The growth was only the size of a pea.
[d] An extensive incision was made, but the patient nearly died from hæmorrhage. Owing to the intimate connection of the growth with the subjacent parts, it could not be removed.
[e] In this case the growth was not larger than a split tare. After the operation, great œdema took place around the wound. The patient also expectorated a quantity of blood and pus, and suffered from high fever. The voice remained hoarse for some time, but was ultimately restored by inhalations.
[f] I have not been able to obtain the *New York Medical Records*, and my information respecting this case has been obtained from Virchow's *Jahresberichte ueber die Fortschritte*, 1870, vol. ii p. 117.

In order to thoroughly weigh the merits of thyrotomy, it is necessary to consider the prospects of the operation : (1st), in relation to the preservation of life ; (2ndly), in relation to the recovery of voice ; and (3rdly), in relation to the immunity from recurrence. Each of these points will now be considered in detail.

(1st.) *In Relation to Life.*—In division of the laryngeal cartilages, there is always some immediate danger, and 9 out of the 28 cases on record, terminated fatally. In one case the patient died from hæmorrhage seven hours after the operation. In another, the patient lived only seven days, death taking place from pleurisy, and metastatic abscesses in the lungs. In a third, fatal pneumonia supervened at the end of two months. In Dr. Cutter's case (No. 19), the patient was almost suffocated during the operation ; and in one of Navratil's cases (No. 23), the hæmorrhage was alarming, and the patient nearly died under the operation, from the quantity of blood which passed down the trachea. In another of Navratil's cases (No. 24) the patient suffered from high fever after the operation, and expectorated a quantity of blood and pus : œdema took place round the wound, and the patient was in a very critical state.

The usual risks attending the ordinary operations for opening the air-passages, are also, of course, present, and tracheitis or bronchitis may supervene. In addition to the immediate danger, there is also the contingent risk of chronic perichondritis at a later period, and, as is well known, disease of the cartilages of the larynx, under these circumstances, is synonymous with laryngeal phthisis. This result has, however, only occurred once in the twenty-eight cases in which thyrotomy has been performed.[1]

In six of the nine fatal cases in the Thyrotomy Table, the disease was cancerous (or semi-malignant), but in at

---

[1] In Brauers' case, the repeated destruction with the actual cautery produced great constitutional disturbance. "Le larynx passa à l'état d'induration squirreuse, une fièvre hectique s'alluma," and the patient slowly sank. In this case the induration was probably due to inflammatory thickening and disease of the cartilages.

least one of these, the patient, who might have lived for months, or even a year or two, died from the immediate effect of the operation, and in the others, with one exception, there is no reason to think that the operation was more beneficial than simple tracheotomy would have been.

Comparing this table with the result of my 93 cases treated through the natural upper orifice of the larynx, it will be seen that the prognosis in thyrotomy is much more unfavourable ; for whilst in 28 cases of thyrotomy 9 patients died, not one of my cases terminated fatally.

Mr. Durham[1] gives a table, which, although it makes the operation of thyrotomy appear to be rather less dangerous (probably by massing together cases of *indirect* laryngotomy with thyrotomy), nevertheless shows how much more successful are operations conducted *per vias naturales.* His table gives 24 cases of external incision into the larynx, with 4 deaths, and 114 cases of operations through the mouth, with only 3 deaths. The operation as regards life, is, indeed, probably much more unfavourable than the Thyrotomy Table would indicate, for many cases, in which the operation was performed, were not at all urgent, and had the operation been practised in necessary cases only, the results would have appeared still less satisfactory. Again, it is highly probable that the operation has been *un*successfully performed in some unpublished cases, and the well-known preference which practitioners have for recording their favourable cases must not be forgotten.

(2ndly.) *In Relation to Voice.*—However carefully thyrotomy is performed, there is always a danger of wounding the vocal cords. In some cases this injury may be avoided, but in others, owing to the rapid ascent and fall of the larynx, which takes place when a foreign body is first introduced into the air-passages, it is impossible to prevent wounding the vocal cords. This *contretemps* is stated to have occurred once—viz., in Dr. Cutter's operation; but though unnoticed, it probably happened more frequently. In the case in which the accident is recorded, the injury

---

[1] *Op. cit.,* p. 584.

was followed by loss of substance. In another case, that of Dr. Rauchfuss, it was necessary, in order to extirpate the growth, to cut away one of the vocal cords.

I do not think, however, that slight injury of the vocal cords is necessarily attended with subsequent loss of function. The aphonia which so often results from thyrotomy, is probably due either to inflammatory changes in the hard or soft tissues, near the commissure of the vocal cords, or to a slight dislocation of one, or both of the cords, at their anterior extremity. If, in the union of the thyroid cartilages, the two halves are brought together, so that one cord is, in the very least degree, above the level of the other, the correlation between the two cords must necessarily be destroyed, and aphonia result. Or again, in bringing the two alæ of the thyroid cartilage together, especially when the connecting crico-thyroid and thyro-hyoid membranes have been divided, the level of the pomum Adami may be altered. Any modification of position at the anterior part of the larynx, of course alters the plane of the vocal cords, and thus affects vocalization. In some cases, the aphonia is probably due to the extensive origin of the growths, and to their intimate incorporation with the subjacent tissues; but that it is due in most cases to the operation itself, is shown by the fact that when thyrotomy is practised for the removal of foreign bodies, the voice is often lost. Out of six cases of this kind collected by Planchon,[1] the voice is only stated to have been retained twice; in one instance there was persistent aphonia, in another slight hoarseness, and in two cases the condition of the voice after the operation is not stated.

Whatever may be the rationale of the phenomena, the very frequent persistence of aphonia or dysphonia after thyrotomy is indisputable. Out of the 28 cases contained in the Thyrotomy Table, the voice was only restored 8 times. Three of the cases, however, proved rapidly fatal, and in 1 the operation had to be abandoned. There remain, therefore, only 24 cases. In 6 of these the patients died in a

---

[1] *Op. cit.*

few months (or in less than 2 years), 4 having remained aphonic, and two having been dysphonic.

Of the remaining 18 patients who survived, 5 suffered from permanent aphonia, 4 from persistent hoarseness, and 8 completely recovered the voice. In 1 case the result is not stated, but as recurrence took place within a fortnight, the voice could not have been improved.

In my 93 cases, in which internal treatment was adopted, the voice was entirely restored in 75 instances, and in 15 there was improvement of the voice. The advantage of laryngoscopic treatment in respect to restoration of function, is so obvious, that it requires no comment.

(3rdly.) *In Relation to Recurrence of Growth.*—It might be expected that extirpation could be more completely effected when the thyroid cartilage is divided, and the larynx thoroughly exposed to view, and that thus recurrence would be less frequent ; but hitherto this does not appear to have been the case.

On analyzing the Thyrotomy Table, it appears that 4 cases rapidly terminated fatally, and therefore gave no time for recurrence.[1] In 6 other cases, the patient died at the end of a few months, and in nearly all of these, recurrence had taken place. They were all, however, of malignant or semi-malignant character, and therefore the tendency to repro-duction was no doubt very great. Of the remaining 18 cases, in one (No. 23) the growth could not be extirpated at all, owing to its close incorporation with the subjacent tissues ; and in two others (Nos. 17 and 18) the neoplasm was incompletely removed. In addition to these cases of incom-plete extirpation, which would be much more numerous if the fatal cases were not eliminated, recurrence took place in two cases (Nos. 19 and 27) in less than a month, and in one case (No. 20) at the end of two years.

In my 93 cases, treated through the fauces, recurrence took place in 6 cases, in which the growth had been previously

---

[1] This includes Ehrmann's case, which is justly considered as a case of recovery with aphonia, the patient having died from typhus ; as, however, she only lived seven months, there was not time to form an opinion as to recurrence.

completely extirpated. In several of these cases the recurrence did not take place till a year or two after evulsion had been effected ; and in most cases the neoplasm occurred in different situations to that which it had previously occupied.

In three cases in which incomplete evulsion [1] was effected, but the symptoms were relieved, the growth after a time underwent further development, and again necessitated treatment. In one case [2] continuous growth has taken place. It will be seen, therefore, that thyrotomy does not even effect such complete evulsion as laryngoscopic treatment.

Removal of Growths by Division of the Thyro-Hyoid Membrane, or Supra-Thyroid Laryngotomy.

Indications for Operation.—This method of treatment is indicated for the removal of large growths situated at the upper orifice of the larynx, which cannot be taken away *per vias naturales.*

History.—This operation, originally proposed at about the same time by Malgaigne [3] and by Vidal [4] de Cassis, was first carried out in the year 1859. The operator was Dr. Prat, a surgeon in the French navy, stationed at that time at Papiete, the capital of Otaheite. The patient, who was the subject of advanced pulmonary phthisis, suffered also from such extreme difficulty of swallowing, that he could scarcely take any food. The dysphagia was due to a growth, which appears to have been situated on the under-surface of the epiglottis ; it could be felt with the finger, but all attempts to seize and remove it through the mouth entirely failed. By operating after the manner recommended by Malgaigne, Dr. Prat easily removed the growth, which was of a compact fibrous tissue and greyish-white colour. No vessels

---

[1] Appendices A and C, Cases 37, 61, 74.
[2] Appendices A and C, Case 59.
[3] The claim to originality is made by Malgaigne in his *Manuel de Médecine opératoire.* Paris, 1871, 7me édition, p. 525.
[4] Velpeau, *Médecine opérat.*

were tied. The wound healed quickly, and the symptoms from which the patient had suffered, disappeared. He died shortly afterwards from phthisis, and at the autopsy no trace of the growth was to be found.[1] Had this difficult, and at that time unique operation, been performed, with the aid of able colleagues, in a first-class European capital, it would have reflected great credit on the practitioner; but when it is recollected that it was done by a naval surgeon in one of the small Polynesian Islands, it is impossible to speak too highly in praise of the boldness and skill of the operator.

In the year 1863 Follin[2] performed a similar operation with complete success. The patient was a young man, aged 21, whose respiration was normal when in the horizontal position, but who could not breathe when standing upright. His symptoms were due to the presence of several fibro-cellular or myxomatous growths, which had rapidly formed, and were thought to be situated on the posterior wall of the larynx. The neoplasms were extirpated, and the patient was entirely cured.

Dr. Follin thus describes his case :—" The operation was performed on the 24th of February, 1863. In order to avoid section of the base of the epiglottis, the incision was made at about 1-8th of an inch above the upper border of the thyroid cartilage. The skin, areolar tissue, internal fibres of the platysma myoides, and the sterno-thyroid and thyro-hyoid muscles, were divided. The superficial incision was about 2¾ inches in length, but in the deeper parts it was much less extensive. No arterial ligature was required, and the venous hæmorrhage was exceedingly slight. The fatty mass in front of the epiglottis was divided, and the laryngeal mucous membrane being divided with scissors, the cavity of the larynx was laid open. This incision, which passed below the base of the epiglottis, led at once to the polypi; and the superficial parts being retracted, about ten excrescences of various size, some as large as Barcelona nuts, were successively removed with great ease by torsion

---

[1] *Gazette des Hôpitaux*, 1859, No. 103, p. 809.
[2] Appendix D, Case 56, and *Archives Générales de Médecine*, Février, 1867.

or slight incisions with scissors." The wound healed rapidly, and the patient left the hospital on the 16th of March, perfectly cured. It should be mentioned in this case, that numerous unsuccessful attempts were made in the first instance to remove the growth through the mouth, but the extreme irritability of the pharynx rendered laryngoscopic treatment impossible.

Method of Procedure.—Transverse incision through the thyro-hyoid membrane should, according to Malgaigne, be made along the lower border of, and parallel with, the hyoid bone, through the skin, superficial fascia, the inner half of the sterno-hyoid muscles, the thyro-hyoid membrane, and the mucous membrane which extends between the base of the tongue and the epiglottis, and forms the glosso-epiglottic ligament. The side of the epiglottis should then be seized and drawn through the wound. The growth can then be removed, according to the circumstances of the case, by bistoury, scissors, or forceps. It will be recollected that Follin divided the thyro-hyoid membrane along the upper border of the thyroid cartilage, that is, rather lower down than advised by Malgaigne, with a view of avoiding the epiglottis ; and as far as I can gather from the report of his case, the incision was carried further outwards than in Prat's case. The latter procedure certainly renders the epiglottis less likely to be wounded, but little immunity is afforded to the valve by making the incision a few centimetres lower down than recommended by Malgaigne. It must also be remembered that the more external the incision is carried, the greater is the danger of wounding important vessels. In any case, the hyoid branch of the thyroid is not unlikely to be wounded, but this is not a matter of any importance.

Comparative Merits of Operation.—Although sub-hyoid laryngotomy is unattended with any considerable danger, either immediate or remote, I do not think that it will find much favour with those skilled in operating with the aid of the laryngeal mirror; for it happens, that those cases which are favourable to the performance of this operation, are just those which, as a rule, can be most easily treated through the mouth.

The two cases in which it has been carried out, are not sufficient for the deduction of any conclusions as to the value of this mode of treatment ; but it appears to me almost certain that in Prat's case the growth might have been removed through the fauces, had the laryngoscope[1] been in use at that time. From the ease with which the growth was removed, it is highly probable that it was situated on the upper or lingual surface of the epiglottis, or, at any rate, on the edge of the valve.

In Follin's case, the polypi were thought to be " attached to the mucous membrane covering the anterior surface and bases of the arytenoid cartilages," but it is much more probable that they grew from the posterior wall of the pharynx. In the report they are stated to have " only covered the posterior half of the glottis ;" and taking their number and size into consideration, it is quite impossible that they could have been situated as described. Had there been room for them in the larynx, they must have given rise to permanent dyspnœa, and must have covered more than the posterior half of the glottis ; on the other hand, if the growths had been attached to the posterior wall of the pharynx, the occurrence of dyspnœa, when the patient stood up, would be explained. In the case of large growths, as already pointed out, the origin is so often concealed, that it is not improbable that a slight error as to the origin of the growth may have been made.

The operation is much less serious than thyrotomy, in relation to life, and is not attended with any risk to the vocal function. In injury of the cartilages which form the framework of the larynx, there is, as has been already pointed out, always the danger of subsequent caries ; but it is well known that injury of the elastic cartilages, though it may cause temporary inconvenience, is unattended with

---

[1] Although the laryngoscope was brought into use in 1859, the year of Prat's operation, it is highly improbable that it had reached the Polynesian Islands at that time.

permanent risk. Not only do we frequently find that patients, recovered from tertiary syphilis, with the mere stump of an epiglottis, can swallow perfectly well ; but it has already been proved, in the celebrated case of Prince Murat,[1] that the epiglottis may be *suddenly* cut away with only temporary inconvenience. Again, most hospital surgeons must have frequently met with extensive suicidal wounds of the thyro-hyoid membrane involving the epiglottis, which have healed rapidly without any bad results. This last fact has been illustrated by some remarkable cases by Künst.[2]

Removal of Growths by Infra-Thyroid Laryngotomy (through the Crico-Thyroid Membrane), or by Tracheotomy.

Indications for Operation.—This operation is recommended for the removal of laryngeal growths situated in the sub-glottic region, as well as for tumours in the upper part of the trachea, when, in such cases, laryngoscopic treatment cannot be carried out.

History.—This mode of eradicating growths was recommended by Professor Czermak in the year 1863 ; but it was first successfully employed two years later by Dr. Burow, senior,[3] of Koenigsberg. In the year 1869, it was carried out, for the second time, by myself.[4]

In Dr. Burow's case the patient was a man, aged 48, who had a growth at the anterior part of the right vocal cord, which blocked up rather less than one-third of the laryngeal cavity, and produced dysphonia and dyspnœa. After repeated unsuccessful attempts to remove the growths through the fauces, Dr. Burow made an opening in the crico-thyroid

---

[1] In this historical case, which occurred at the battle of Aboukir, half of the epiglottis was carried away by a musket-ball. Under Baron Larrey's treatment the patient recovered. Another similar case occurred in the same campaign, with an equally fortunate result.—(Larrey, *Clinique chirurg.*, tom. ii. p. 142 ; *Relation chirurg. de l'Armée d'Orient*, p. 286, quoted by Ryland.)

[2] *Eröff. der oberst. Luftwege.* Leipzig, 1864, p. 45.

[3] *Deutsche Klinik.* vol. xvii. p. 165 ; and Appendix D, Case 110.

[4] Appendices A and C, Case 81.

membrane, and succeeded with a hook in removing almost the entire growth. No canula was inserted. The patient's breathing became immediately easy, and the voice clearer. When the patient was discharged, his voice, though much improved, was not quite resonant. As a proof of the very slight constitutional disturbance produced by the operation, Dr. Burow states that his patient left the hospital an hour after the operation, and that he was never confined to his bed, nor even to the house.

In my own case, which is hereafter related in detail, the patient, a woman, aged 51, was admitted into the Hospital for Diseases of the Throat, on May 27th, 1869, on account of extreme dyspnœa. On laryngoscopic examination, it was seen that there was paralysis of the abductors of both vocal cords ; so that the glottis was reduced to the narrowest chink, and no view could be obtained of the parts below. Laryngotomy was at once performed, on account of the dyspnœa ; and it was not till a week later, that, on the first laryngoscopic examination after the operation, a growth was discovered to be the mechanical cause of the paralysis. The neoplasm was easily removed with curved forceps, introduced through the wound. The tube was removed six months later. The respiration remained normal and perfectly natural eighteen months after treatment.

At the time, when this patient was under my care, I was not acquainted with Dr. Burow's interesting case.

Method of Procedure.—An incision should be made as in ordinary (crico-thyroid) laryngotomy, but the crico-thyroid opening should be carefully dissected out, and all the membrane, muscle, and superficial parts removed, so that nothing is left but the two cartilages surrounding the opening, and a canula inserted a few days before evulsion is attempted. When all disposition to hæmorrhagic oozing has ceased, and all tenderness disappeared, the canula should be taken out, the chin thrown well back, so as to enlarge the crico-thyroid space as much as possible, and a careful examination made with one of Neudörfer's infra-glottic mirrors, to ascertain the exact origin of the growth. The mirror must then be dispensed with, and the growth removed with short tube-forceps.

This operation can only be performed where the crico-thyroid membrane is of average size, and if there is not room to effect removal, tracheotomy should be performed in the first instance instead of laryngotomy. The steps of the operation are almost the same as in (crico-thyroid) laryngotomy. When the patient has recovered from the tracheotomy, that is to say, a few days after the operation, the canula should be removed, and an attempt made to extirpate the growth. In carrying out the operations, the two sides of the windpipe require to be held back with retractors, in order that instruments may be conveniently passed into the larynx.

The patient should continue to wear the canula for a few months, or, at any rate, for a few weeks, in case eradication be incomplete, or recurrence take place. Had Dr. Burow taken this precaution, the result of his case would have been even more satisfactory.

Comparative Merits of Operation.—It is very remarkable that this operation has not been more frequently performed. In my 100 tabulated cases the growth was situated ten times below the vocal cords. In three of these cases evulsion was effected with great difficulty, and only after several months' close attendance. In two of them, indeed, the success was incomplete, a small portion of the growth remaining, and the voice being still a little hoarse. In the cases tabulated in Appendix D, the growths were in 15 cases situated below the vocal cords. In five of these thyrotomy was performed, and in two the less serious step of tracheotomy was necessitated, before the growth could be removed through the fauces. There is little reason to doubt that, in many, if not all, of these cases, had an opening been made in the crico-thyroid membrane, the treatment might have been reduced to a few days, or at the most a few weeks, instead of lasting, as it did, several months, and that it is possible the results would have been altogether more satisfactory. Nevertheless, it must not be forgotten that many patients would prefer a tedious operation, carried out *per vias naturales*, to a direct incision through the neck, however slight the operation may be. In one of my cases the operation through the crico-thyroid membrane was suggested by me, but declined by the patient. Although there is a much greater prospect of success if a canula is employed, as advised above, the contingent danger of subsequent bronchitis is probably greatly reduced by dispensing with it, after the manner of Dr. Burow.

THE COMBINED METHOD OF REMOVING GROWTHS . (TRACHEOTOMY OR LARYNGOTOMY BEING FIRST PERFORMED, AND THE GROWTH BEING SUBSEQUENTLY REMOVED BY LARYNGOSCOPY).

This method of treatment may be advantageously carried out in those cases in which the size of the growth causes immediate danger to life, and where laryngoscopic treatment gives rise to serious dyspnœa.

It has already been remarked that operative procedures on large growths are apt to give rise to inflammatory tumefaction of the mucous membrane and to spasm of the glottis; and in three cases which have come under my notice, partial evulsion, or attempts at evulsion, through the upper orifice, have precipitated tracheotomy. In these cases it would perhaps have been as well to have opened the windpipe in the first instance, and only subsequently to have attempted removal through the mouth. Where the combined method of treatment is adopted, the patient should be allowed a few days' rest after tracheotomy has been performed, and only when thoroughly recovered from the operation, should eradication be attempted, with the aid of the laryngoscope. The practitioner will be guided by the observations already made under the sub-sections on Mechanical Treatment and Chemical Treatment, as to the best method of removing or destroying the neoplasm, and only when the growth is completely eradicated should the canula be removed.

This method of treatment holds a middle place between laryngoscopic and extra-laryngeal treatment, and while on the one hand the practitioner should always first endeavour to remove a growth through the fauces, it is better, if he does not succeed, to adopt the operation of tracheotomy as a safeguard, and then to pursue laryngoscopic treatment. If then he is still unable to remove the growth through the mouth, he may proceed to the more serious operation of thyrotomy. In my own practice I have successfully employed the combined treatment in two cases.[1] In two others,[2] it became necessary to perform tracheotomy, on account of the difficulty of removing the growth *per vias naturales*, and the respiration being relieved by this operation, no further treatment of the neoplasm was permitted.

The combined method has been frequently pursued on the Continent and in America, and 11 cases are contained in Appendix D. Of these, 5 were cured, and in the remaining 6 the dyspnœa was entirely relieved and the voice improved. It will thus be seen that the combined method offers very satisfactory prospects.

[1] Appendices A and C, Cases 74 and 88.
[2] Appendices A and C, Cases 36 and 45.

APPENDIX A.

— *a*

REPORTS OF ONE HUNDRED CONSECUTIVE CASES
TREATED BY THE AUTHOR.

15

APPENDIX A.

— o —

CASE I.—(*Papillomatous*) *Growth on the under Surface of Epiglottis; Irregular Ulceration of the edges of the Vocal Cords; Treatment by Caustic Solutions; Improvement.*

Mr. W., of Sligo, a bookseller, aged about 45, consulted me in June, 1862, for loss of voice, which had existed since November, 1859. On examining the throat with the laryngoscope, the vocal cords were seen to be of a dirty grey colour, and in a highly disorganized state,—their edges being serrated in a very peculiar manner; the rough toothlike processes of one vocal cord fitted into corresponding depressions in the edge of its fellow. In the middle of the under surface, and near the edge of the epiglottis, was a small round excrescence (Fig. 9). There was nothing of a

Fig. 9.

syphilitic character in this case; and the diseased condition of the larynx seems to have originated in a severe and prolonged catarrh. In this case, I enjoyed the advantage of a consultation with Professor Czermak. He agreed with me in thinking it a very unfavourable one for treatment,—at least as regarded the condition of the vocal cords. He also concurred with me in recommending the local application of strong solutions of nitrate of silver. Neither of us, however, was at all sanguine as to the effect it might produce. I applied this remedy to the interior of the larynx eight or nine times, but without any apparent effect. Mr. W.

then returned to Ireland, and the same treatment was continued by Dr. Wood, of Sligo. After many months' treatment, the whisper was replaced by a very gruff voice; and when Mr. W. came over to consult me in September, 1863, the edges of the vocal cords were much more even,—the right cord being almost smooth, and the voice, though rather hoarse, was distinctly phonetic. The small warty growth had slightly diminished in size.—*The Use of the Laryngoscope,* second edition, 1866, page 86, and *Jacksonian Prize Essay, On Diseases of the Larynx, &c.,* 1863, Case 20. MS. in Library of the Royal College of Surgeons.

CASE II.—(*Papillomatous*) *Excrescences on the Right Ary-epiglottic Fold and Ventricular Band; Treatment by Caustic Solutions; Improvement.*

C. A., æt. 42, from Diss, in Norfolk, applied at the Hospital for Diseases of the Throat, in April, 1863. This patient was the mother of a large healthy family. She had suffered from loss of voice and a constant inclination to clear the throat for two years, but was otherwise quite well; she attributed the aphonia to having taken cold. With the laryngoscope, the aphonia was seen to depend on the presence of numerous small warty growths, situated on the right ary-epiglottic fold and ventricular band. There was also slight congestion of the vocal cords (Fig. 10). Under

Fig. 10.

the use of solutions of nitrate of silver (ℨij ad ℨj), applied very frequently for some weeks, this patient became able to speak in a fairly loud, though still rather harsh, voice. When the patient was obliged to return to the country, the greater part of the excrescence had been destroyed, but a small portion still remained on the ventricular band.—*The Use of the Laryngoscope,* second edition, page 87, and *Jacksonian Prize Essay,* 1863, Case 19.

CASE III.—*Five large Papillomatous Excrescences in the Interior of the Larynx; Treatment by Evulsion and Escharotics; Cure.*

William W., æt. 44, carpenter, applied to me April 10th, 1863, on account of loss of voice. He stated that his general health was very good, but that three years ago he had caught a cold and bad sore throat, and since then he had not been able to speak a word out loud. At Christmas his breathing was much affected, and he thought he should have been suffocated; but the attack passed off, and he said that, with the exception of not being able to speak out loud, he was now quite well. He had never had syphilis. On making a laryngoscopic examination, the laryngeal mucous membrane, above and below the vocal cords, was seen to be covered with dark, reddish, spongy excrescences (Fig. 11, and Plate II. fig. 3). One was situated on the right side of the under surface of the epiglottis; another involved the whole right ventricular band; a third covered the whole of the right vocal cord; a fourth occupied half of the left ventricular band; and a fifth the anterior half of the left vocal cord. Below the right vocal cord a number of smaller excrescences were also seen extending down into the trachea.

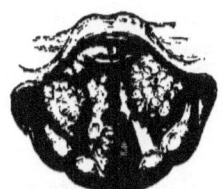

Fig. 11.

This case was seen by Drs. Czermak, Frodsham, George Johnson, Wahltuch, and others. With my laryngeal forceps, I succeeded, in a number of sittings, in removing, in small fragments, the whole of the four upper excrescences. This included the one seated on the left vocal cord. These fragments were kindly examined for me by Dr. Andrew Clark. He "found them to consist of numerous yellowish, hard, nodular, or warty-looking particles. Under the microscope, some of these masses consisted entirely of enlarged racemose glands, clothed with many layers of epithelium, the outermost layer of which was in a state of partial desquamation. A few of the papillæ were either quite hollow, or had contained fluid." He

regarded the case as one of "Granular Wart." The small particles
which were torn away with the forceps produced so little effect on
the bulk of the large growth on the right vocal cord, that I was
induced to try the effect of escharotics. Nitric acid and chromic
acid were both applied several times with decided advantage ; but
the greatest benefit resulted from the employment of a mixture of
caustic soda and lime. In August, the patient recovered a loud and
tolerably clear voice. This patient remained under observation
until November, 1866, when all the growth had been removed and
the voice completely restored.—*The Use of the Laryngoscope*, second
edition, page 121, and *Jacksonian Prize Essay*, 1863, Case 21.

This patient returned to the Hospital, March 7th, 1870, stating
that since his discharge in 1863, his voice had remained perfectly
clear till about four months previous to his present application. It
was now gruff and husky.

On laryngoscopic examination, the larynx was seen to be almost
entirely occupied by two symmetrical growths of a pink colour and
irregular surface, proceeding from each ventricular band. So large
were the growths, that, in attempted phonation, they met in the
median line, and only the posterior half of the vocal cords was then
visible.

The case was seen before treatment by Dr. Carpenter, F.R.S.,
and during his attendance by Dr. Stage, of Copenhagen, and Mr.
Keene. In four sittings, the larynx was entirely cleared, and the
voice restored to its normal tone.

CASE IV.—(*Papillomatous*) *Excrescences on and beneath both Vocal
Cords ; Treatment by Evulsion ; Cure.*

Mrs. A., æt. 35, the wife of a mechanic, applied at the Hospital
for Diseases of the Throat in April, 1863, though, in consequence of
my absence from town, she did not come under my care till the
following month. I had previously (in December, 1862) seen the
patient, at Mr. Maunder's request, in conjunction with Dr. Gibb ;
and the latter author has referred to the case (*Diseases of the Throat*,
page 156. London, 1864) and given a rough sketch of the laryngoscopic appearance. The patient stated that she caught cold in
1859, was very hoarse for two years, and that in 1861 her voice
had become quite suppressed. For the last two years she had
always spoken in a whisper. There was no history nor symptom

of syphilis or phthisis. With the laryngoscope, both vocal cords were seen to be of a dirty greyish colour, and in an irregular papillomatous condition : the appearance is shown in Fig. 12.

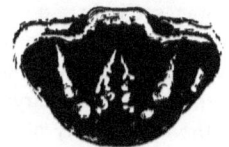

Fig. 12.

Subsequently I discovered two growths,—one below each vocal cord. As the diseased condition of the cords was so general, and the growths on the cords were so imperfectly developed, I thought that it would be best to treat the case with caustics. Strong solutions of nitrate of silver were accordingly applied, but they produced so much dyspnœa, that the treatment was obliged to be discontinued. Under these circumstances, I tried to use forceps; but the patient being unable to open her mouth widely, the laryngeal aperture being exceedingly small, and the growths on the vocal cords most minute, great difficulty was experienced, and it was only after repeated failures that I ultimately succeeded in clearing the vocal cords of the warty growths which covered them. The growths below the cords, which afterwards became distinctly visible, being of larger size, were removed with much less difficulty. A month after the removal of the last growth, the patient's voice was completely restored. I have not seen her now for some time, but I received a note (dated October 31st, 1864) from Mr. Brown, of Finsbury Circus, who sent the patient to me, in which he says, " I called on Mrs. A. this evening, and am pleased to find her voice is entirely restored by your treatment."—*The Use of the Laryngoscope*, second edition, page 124, and *Jacksonian Prize Essay*, 1863, Case 26.

CASE V.—*Papillomatous Growth on the Left Vocal Cord ; Treatment by Evulsion ; Cure.*

Henry R., æt. 45, a gas-fitter, applied at the Hospital for Diseases of the Throat, May 1st, 1863, on account of loss of voice of nine years' standing. He stated that he had attended at various metropolitan hospitals, and had lately been at the

Brompton Hospital. On examining his throat with the laryngo-scope, a small round excrescence, about the size of a pea, was seen on the left vocal cord. The warty growth was situated on the free edge, and exactly in the middle of the cord, and on attempted phonation it was seen that, owing to the projection of the growth, the cords could not become approximated. On the right cord, exactly opposite to the wart on the left cord, there was a distinct round indentation. The laryngoscopic appearance is seen in Fig. 13. I had the opportunity of exhibiting this patient to

Fig. 13.

Drs. Czermak; Wahltuch, and others. There was some difficulty in removing this growth, owing to its small size, and the more than usual awkwardness of the patient, and it was not till the fourth sitting that it was successfully seized and removed. It may be remarked, that solid nitrate of silver had been previously, repeatedly applied, but without benefit. Dr. Andrew Clark examined micro-scopically the portions removed with the forceps, and the following was his report:—" The growth was found to consist of two sets of particles, one membranous, the other warty or obscurely papilliform. The membranous portions consisted of from twenty to thirty layers of scaly epithelium, surrounded and penetrated by a confervoid growth. The epithelial cells composing the layers were polygonal, flattened, nucleated, and easily affected by weak alkalis and acids. The nucleus of each cell was oval, abruptly defined, rather large in proportion to the containing cell, in most cases surrounded by a clear halo, and in some showing signs of division. The papillary portions consisted of simple outgrowths of nucleated connective tissue and rudely-formed blood-vessels, clothed with numerous layers of scaly epithelium, similar to those already described. Some of the papillæ exhibited large vacuoles or spaces filled with colloid matter, which, in one or two instances, had burst through the cover-ing epithelium." Dr. Clark considered the tumour to be a true wart. Immediately after the operation the patient spat up a few teaspoonfuls of blood, and the same day he was able to sound his voice. The next day he complained of a feeling of great soreness, and there was so much involuntary objection to a laryngoscopic examination, that I was unable to see exactly how the wound looked. Nine days later, however, the mucous membrane over the left vocal cord, where the growth had been, looked rather puckered, and the depression on the right cord was still visible. At the end of a month the voice was perfect, and all morbid appearance in the larynx, including the little pit on the edge of the right cord, had completely disappeared.—*The Use of the Laryngoscope*, second edition, page 125, and *Jacksonian Prize Essay*, 1863, Case 22.

CASE VI.—(*Papillomatous*) *Growths on both Vocal Cords ; Treatment by Evulsion ; Cure.*

William J., æt. 40, applied to me in May, 1863, on account of hoarseness of five years' standing. His general health was good, but fifteen years before he had a primary venereal sore. He had never suffered from any secondary symptoms. The voice was harsh, but not suppressed ; and with the laryngoscope a large, thin, flat, membranous growth was seen to project horizontally from each vocal cord, and to meet in the centre. On account of the pendulous condition of the epiglottis, it was difficult to get an extensive view of the larynx, and consequently the growths could not be seen in their entirety. The appearance is shown in Fig. 14. The smallness

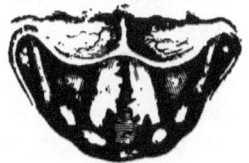

Fig. 14.

of the laryngeal aperture was still more inconvenient in operating ; and it was only after several unsuccessful attempts that I managed to remove a small portion of the growth on the right vocal cord. Under these circumstances, I endeavoured to divide the left growth through its base, with my laryngeal lancet. After the operation, the patient left me, but soon returned, spitting up considerable quantities of blood. On examination with the laryngoscope, the mucous

16

membrane was seen to be covered with blood; but the exact source of the hæmorrhage could not be ascertained. I applied a strong solution of perchloride of iron to the interior of the larynx, and directed the patient to suck ice. The hæmorrhage, however, which continued for some time—to an extent that was really alarming,—was ultimately arrested by.the patient gargling with, and swallowing, a saturated solution of tannin. The first mouthful of the tannin that was swallowed stopped the bleeding entirely. A day or two after the operation, a careful examination of the larynx was made both by Dr. George Johnson and myself, but we were neither of us able to ascertain the source of the hæmorrhage. I have since removed several fragments by using the horizontal blades of my tube-forceps, and the patient's voice is now clear; he still complains, however, of a slight tickling in the throat.—*The Use of the Laryngoscope*, second edition, page 126, and *Jacksonian Prize Essay*, 1863, Case 23.

CASE VII.—*Fibrous Growth above the Anterior Commissure; Treatment by Evulsion; Cure.*

Morris B., æt. 41, shoemaker, and formerly singer, applied at the Hospital for Diseases of the Throat, August 20th, 1863. He stated that he had been extremely hoarse for seven years, but had never suffered from complete loss of voice. He had been affected with primary syphilis when he was sixteen. A physician had recommended him to have his uvula removed, but the operation had not improved his voice. A laryngoscopic examination showed that there was a yellowish-pink growth, about the size of a small bean, just above the anterior insertion of the vocal cords. It was moveable (and therefore probably pedunculated), but the base was hidden by the tumour, and therefore its exact origin could not be ascertained. When the glottis was closed, the growth rested on the extremities of both the

Fig. 15.

cords, sometimes, however, lying more on the right, and sometimes on the left cord. The appearance is shown in Fig. 15.

Aug. 21.—I had the advantage of a consultation with Dr. George Johnson and Mr. Mason, who entirely concurred in my diagnosis.

Aug. 24.—In the presence of these two gentlemen, I removed the excrescence with my tube-forceps. The growth was successfully seized at the first trial, and all of it, except a small portion of its base, was brought away. After the operation, we examined the patient with the mirror, and the base of the growth covered with blood was indistinctly seen. I was disposed to remove this small remaining fragment, but after a consultation, it was thought better to leave it alone, under the idea that it would probably wither away.

Immediately after the operation, Dr. Johnson thought he noticed an improvement in the voice.

Aug. 26.—There being still a small portion of the base of the growth remaining, I removed it with tube-forceps. The patient was completely cured, and at the end of a fortnight he spoke perfectly well.

"The morbid growths," according to Dr. Andrew Clark, "consisted of three or four minute, shapeless pieces of yellowish colour, streaked with red, and of a horny consistence. On account of their hardness, their structure could not be very easily determined. On the free surface, however, were several layers of thin, scaly epithelium, few of the elements of which exhibited any nuclei. In fact, but for the absence of cholesterine, the cell-elements might have been most readily mistaken for those of cholesteatoma. Beneath the epithelial coverings were minute extravasations of blood, and amorphous masses of a coagulated proteine compound." Though in this case the proteine compound had not developed fibres, the case was regarded by Dr. Clark as one of commencing "Fibro-epithelial Growth."—*The Use of the Laryngoscope*, second edition, page 128, and *Jacksonian Prize Essay*, 1863, Case 25.

CASE VIII.—*(Papillomatous) Growths on the Epiglottis and Right Ventricular Band; Treatment by Evulsion; Cure; Subsequent Recurrence; Treatment by the same method; and Complete Restoration of Voice.*

George T., æt. 26, a sailor, applied at the Hospital for Diseases of the Throat, January 28th, 1864, on account of slight dysphagia, from which he had suffered for six months. He had lately returned from Hobart Town, and on his voyage round Cape Horn, had been exposed to severe cold during his night watches; since that time he had suffered from hoarseness. On laryngoscopic examination, two small excrescences were seen to be sprouting from the under surface of the epiglottis, and one large foliate growth from the right ventricular band (Fig. 16).

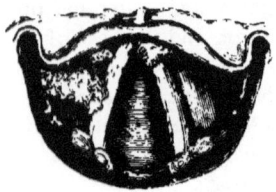

Fig. 16.

These growths were removed, without much difficulty, in five sittings, partly with common laryngeal forceps, and partly with tube-forceps; and both the dysphagia and hoarseness were entirely relieved. The patient was seen both before and after treatment by my colleague, Mr. George Evans.

The man applied again at the Hospital, February 21st, 1866, stating that he had been three voyages since his former attendance, and that it was only during the last that he had suffered from his old symptoms. For a month his voice had been completely suppressed, and he had been hardly able to do any work of an active character, on account of attacks of shortness of breath, amounting almost to suffocation, which supervened on the slightest exertion.

On laryngoscopic examination, irregular warty excrescences were discovered beneath the anterior commissure of the vocal cords. There was no appearance of recurrence in the situation of the former growths.

The whole of the new formations were removed with tube-forceps, after an attendance of three months. The case was seen during this second course of treatment by Drs. Tatham, Thurgar, and Wilkins. On microscopic examination the growths were found to be of a papillary structure, in which the connective-tissue element could not be detected. The laminæ of epithelium were very abundant, twenty-eight layers having been counted in one part.

CASE IX.—*Benign Epithelial Growth on the Right Vocal Cord ; Partial Evulsion ; Great Improvement.*

Mrs. S., æt. 45, a married lady, residing in Manchester, consulted me in January, 1864, on account of great dyspnœa and loss of voice. The symptoms had been coming on during the previous three years, but the dyspnœa had become much more severe during the last six months ; and whilst for the first two years she had been able to speak in a hoarse voice, during the last twelve months she could only whisper.

On making a laryngoscopic examination, a cauliflower-looking growth, about the size of a large pea, was found occupying the anterior third of the right vocal cord.

An attempt was made with tube-forceps to remove the growth, but it caused so much spasm that it was found necessary to desist.

A week later, a large piece—rather more than half the entire growth—was removed.

On the 14th of February another smaller piece was removed in the same way. After this, there was no further dyspnœa, and the voice very much improved.

The portions removed were examined by Dr. Andrew Clark, and found to consist almost entirely of epithelial scales, in various stages of growth, with a very small amount of connective tissue at what appeared to be the base of the tumour.

CASE X.—*(Papillomatous) Growths on both Vocal Cords ; Treatment by Evulsion ; Cure.*

Miss Mary B., æt. 30, was sent to me by Mr. Parsons, of Bridgewater, April 7th, 1864. This patient lived in London, and after she had been suffering from loss of voice for some months, a distinguished physician recommended "change of air to her native place." On arriving there (Bridgewater) she was recommended to return back to London to see me, and the laryngoscope at once revealed the cause of the hoarseness.

A small growth was seen on the right vocal cord, and afterwards, when the patient had been examined once or twice, another growth was perceived on the left cord, near to its anterior insertion. The

appearance is shown in Fig. 17. The history of the case seemed
to show that these growths originated in chronic laryngitis. After

Fig. 17.

twenty attempts, only four of which were successful, the growths
were entirely removed with the forceps. After the removal of the
warts from the vocal cords, a small growth was seen lower down ;
but as the voice was restored, no further treatment was adopted.—
The Use of the Laryngoscope, second edition, page 130.

CASE XI.—*Papillary Growths above and below the Anterior Commissure of the Vocal Cords ; Treatment by Evulsion ; Cure.*

Caroline M., æt. 4, who for two years had been suffering from
loss of voice, stridulous breathing, and occasional attacks of suffocation, was placed under my care, at the Hospital for Diseases of
the Throat, November 9th, 1864. The symptoms had come on
two years previously, but had lately become much aggravated.

On examining the patient with the laryngoscope, an oblong
tumour, about three-eighths of an inch long, and a quarter of an
inch broad, was seen to be attached just above the interior insertion of the vocal cords.

A consultation having taken place with Mr. Mason, who also
examined the child with the laryngoscope, I removed the growth
with the tube-forceps. After its evulsion, numerous excrescences
were seen below the anterior insertion of the vocal cords. These,
after several operations, were also removed.

The respiration became easy as soon as the large growth was
removed, but it was not until the smaller ones were eradicated that
the voice was restored. The portions removed were found to be
of simple papillary structure.—*Transactions of the Pathological
Society*, vol. xvi. page 38 ; also, *The Use of the Laryngoscope*, second
edition, page 133.

CASE XII.—*Numerous Papillomatous Growths on the Vocal Cords and Ventricular Bands; Treatment by Evulsion; Cure.*

Ellen B., æt. 6, was brought to me at the Hospital for Diseases of the Throat, November 20th, 1864, on account of loss of voice, which had existed for two years. The patient had been under the care of several medical practitioners, and among others, under that of Dr. Martyn, of Knightsbridge, who had used the laryngoscope and recognized the nature of the disease. Finding numerous excrescences on the vocal cords and ventricular bands (Fig. 18), before operating I had a consultation with Mr. Mason.

Fig. 18.

Notwithstanding the early age of the patient, the growths were easily removed with tube-forceps, and the voice immediately became *phonetic;* for a long time, however, the little girl remained slightly hoarse. On microscopic examination, the growths were found to consist of several layers of squamous epithelium, a few papillæ, and some enlarged racemose glands.—*Transactions of the Pathological Society*, vol. xvi. page 39 ; and *The Use of the Laryngoscope*, second edition, page 133.

CASE XIII.—*Small (Papillomatous) Growth on the Cartilaginous portion of the Glottis on Right Side; Treatment by Evulsion; Cure.*

Harriett H., a fish-hawker, æt. 28, applied at the Hospital for Diseases of the Throat, in January, 1865, on account of a constant tickling in the throat, which caused her great uneasiness, and even anxiety. She had a frequent inclination to clear the throat ; but, notwithstanding that her occupation required the constant use of her voice at a high pitch, she was not at all hoarse.

On making an examination, the pharynx was seen to be considerably relaxed, and the uvula somewhat elongated. Without further examination, this might have seemed sufficient to account for her symptoms ; but with the laryngoscope, a small warty growth was at once discovered on the posterior, or cartilaginous portion of the right vocal cord (Fig. 19). The growth appeared to prevent the approximation of the cartilaginous portion of the glottis, but did

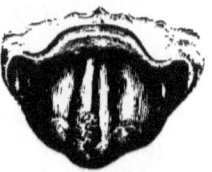

Fig. 19.

not at all interfere with the production of sound.   The patient was treated for pharyngeal relaxation, and a portion of the uvula was removed ; but at the end of six weeks, the troublesome symptoms not being at all relieved, treatment was directed to the removal of the growth by the common antero-posterior forceps.   After a few sittings the little wart was removed in the presence of Dr. Mill Frodsham, and the cough entirely ceased a week later.   The patient was seen in July, 1865, when she still remained quite well.

This case is one of unusual interest, both in its clinical and its physiological bearings.   It shows, in the first place, that a combination of symptoms indicative of pharyngeal disease may depend on a morbid condition of the larynx ; and, secondly, it demonstrates conclusively, that the closure of the posterior part of the glottis is not essential to the production of sound.

CASE XIV.—*Numerous small Benign Epithelial Excrescences on both Vocal Cords and the Right Ventricular Band ; Partial Evulsion ; Improvement.*

J. C., æt. 46, hawker, applied at the Hospital for Diseases of the Throat in the latter part of January, 1865, on account of aphonia. He had been hoarse for eighteen months, but the voice had only been entirely suppressed for the last seven weeks.   He attributed his affection to his having been obliged to exert his voice while suffering from cold.

On laryngoscopic examination, the right ventricular band and both vocal cords were seen to be covered with short, sessile, foliate, excrescences of a pinkish colour.   The larynx was generally much congested.   Inhalations of compound tincture of benzoin were ordered ; and at the third visit, on the 14th of February, several small portions of the growths were removed with common laryngeal forceps. Microscopic examination, kindly conducted by Dr. Andrew Clark, showed the growths to be of a "simple warty character." No papillary structure could be found. On two subsequent occasions small portions were removed, and a fairly useful voice was restored. Content with this improvement, the patient discontinued his visits. When last seen, a small portion of growth still remained on the left vocal cord.

CASE XV.—*Papillary Growth on the Right Vocal Cord; Treatment by Evulsion; Cure.*

Conway C., æt. 12, from Gosport, was brought to me in January, 1865, on account of loss of voice and dyspnœa. When five years old the boy had suffered from croup after measles, and since then had not been able to speak out loud. For the last eighteen months he had been short of breath, having been quite unable to play at any games, and several times during the last year he had "seemed as if he would be strangled." On making a laryngoscopic examination, a growth was seen attached to the right vocal cord and beneath the anterior commissure. The appearance is shown in the annexed woodcut (Fig. 20). On February 3rd Mr. Mason kindly made a very

Fig. 20.

careful laryngoscopic examination and drawing of the case. After a great many attempts I ultimately succeeded with my tube-forceps in removing the whole of the growth ; and in the following June, to use Mr. Mason's expression, there was nothing to see but "a slightly uneven condition of the cords :" this afterwards passed away. Dr. Andrew Clark described the specimens which were brought before the Pathological Society, December 19, 1865, as being " examples of simple papillary, warty, or cauliflower growths." The voice was completely restored.—*The Use of the Laryngoscope,* second edition, page 133.

17

CASE XVI.—*Fibrous Growth on the Posterior Wall of the Larynx; Treatment by Evulsion; Recovery of Speaking Voice, but not of Power of Singing.*

George S., æt. 37, a vocalist, applied at the Hospital for Diseases of the Throat, May 10th, 1865, on account of hoarseness, which had existed for two years and a half, but had not much increased since its first appearance. There had been no cough nor other symptom.

On laryngoscopic examination, a large pyramidal growth, of a dark-red colour, was seen projecting from the posterior wall of the larynx, between the arytenoid cartilages, into the area of the glottis. The vocal cords were much congested (Fig. 21).

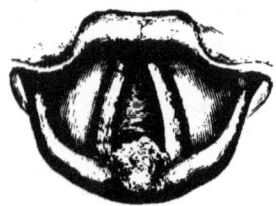

Fig. 21.

The growth was easily removed with tube-forceps at the first attempt, in the presence of Drs. Dale, Frodsham, and others. A month later the patient's speaking voice was perfectly restored, but the power of singing had not returned three months after the removal of the growth. There also still remained considerable congestion of the mucous membrane of the larynx. On microscopic examination the growth was seen to be a true fibroma.

CASE XVII.—*Large Benign Epithelial Growth attached to the Left Vocal Cord; Treatment by Evulsion; Cure.*

Eliza P., æt. 31, a stout healthy-looking woman, from Gravesend, was sent to me in June, 1865, by Mr. John A. Kingdon. She stated that in the winter of 1858-59 she had a bad cough and cold, and that the hoarseness which came on at that time passed in a few months into complete loss of voice; since then she had not been able to speak a word out loud. In 1860 she was an in-patient in a provincial hospital, and there shower-baths were used—but in vain—to restore the voice. In the following year she was admitted into one of the metropolitan hospitals, and here the treatment consisted in blistering the neck, twenty-seven blisters having been applied consecutively; afterwards iodine, mustard poultices, and turpentine stupes were used, and general remedies (quinine, iron, &c.), but all without effect. She further stated that she always suffered now from shortness of breath, and that lately she had had two attacks of great difficulty of breathing, which had lasted for several days. On making a laryngoscopic examination, an irregular lobulated growth, about the size of a sparrow's egg, was seen to be attached to the entire length of the left vocal cord; it projected up into the laryngeal cavity, and across the glottis. At the patient's second visit an attempt was made to seize the growth with tube-forceps, and on the first trial a large piece was seized and brought away. On several other occasions fragments were removed, but the attempts to seize the tumour were frequently quite unsuccessful. On account of the distance at which the patient lived, she was not able to attend at all regularly; and long intervals often intervened between her visits. Accordingly, it was not till March 7, 1866, that I succeeded with a pair of ordinary forceps (opening in the antero-posterior direction) in completely clearing the larynx. Dr. Pratt was present on this occasion, and made a laryngoscopic examination both before and after the removal of the last piece. The patient attended twice at the hospital afterwards, and the larynx was seen to be perfectly healthy. The voice was clear and natural. A portion of the growth examined by Dr. Andrew Clark was pronounced to be "an ordinary warty growth."—*The Use of the Laryngoscope*, second edition, page 134.

CASE XVIII.—*Papillary Growth on the Right Vocal Cord; Dysphonia; Treatment by Evulsion; Cure.*

George D., æt. 47, an engine-driver, applied at the Hospital for Diseases of the Throat, July 5th, 1865, on account of hoarseness. This symptom had come on gradually for about a year; and when he applied, his voice was so hoarse as to be barely phonetic. On laryngoscopic examination, a pink cauliflower-like excrescence was seen occupying the middle third of the edge of the right vocal cord (Fig. 22, and Plate II. fig. 2). The whole of the growth was removed in two operations, but the voice was not completely restored till December, when the patient was examined, and no trace of the growth could be found. In this case, the larynx was examined,

Fig. 22.

both before and after operation, by my colleague, Mr. George Evans. The growth was found, on microscopic examination, to consist of branching papillæ.

CASE XIX.—(*Fibroma*) *on the Epiglottis; Evulsion; Negative Result.*

J. B., æt. 34, a mechanic, came from the neighbourhood of Leeds, to consult me in July, 1865, on account of hoarseness, of two years' standing. During the last few weeks caustic solutions had been applied to the larynx by a local practitioner, but without benefit. On laryngoscopic examination, a polypoid excrescence, of pink colour and smooth surface, was seen covering the anterior third of the right vocal cord, though growing from the lower part of the epiglottis, by a broad base. From its smooth appearance, it was judged to be a fibroma (Fig. 23). Two attempts were made to

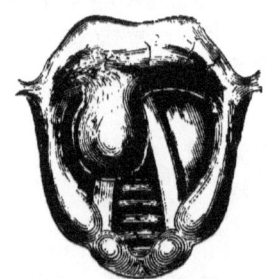

Fig. 23.

remove the growth with the tube-forceps ; but the patient was so nervous, and his throat so irritable, that the growth could not be seized. The patient left London again the same day, and was not seen subsequently.

CASE XX.—(*Papillomata*) *on both Vocal Cords; almost complete Evulsion; Great Improvement.*

John E., æt. 35, labourer, came from Walton-on-Thames to the Hospital for Diseases of the Throat, September 16th, 1865. The only symptom complained of was loss of voice. Hoarseness had existed for two years, and had gradually increased, until, for the last few months, the voice had been entirely suppressed. On laryngoscopic examination, a small irregular growth was seen occupying the anterior two-thirds of the right vocal cord, and another larger one on the posterior two-thirds of the left vocal cord (Fig. 24).

Fig. 24.

The latter growth projected within the larynx, so as to very much diminish the size of the glottis. Notwithstanding this, the patient had never suffered from dyspnœa, nor had his general health been at all impaired. The whole of the growth on the left vocal cord was removed, and a large portion of that on the right cord was removed in the presence of Dr. Baxter, and Dr. Clark of Melbourne. A small piece of the latter, however, remained near the anterior insertion of the cord. The voice, although still gruff, was phonetic and powerful, and the patient discontinued attendance.

CASE XXI.—*Two very Small Papillomata on the Right Vocal Cord; Treatment by Evulsion; Cure.*

Samuel J., æt. 36, tailor, applied at the Hospital for Diseases of the Throat, in December, 1865, on account of dyspnœa and hoarseness. The attacks of shortness of breath came on suddenly, and generally at night; but they sometimes occurred during the day. The first attack had come on eighteen months previously, and the second one six months later. During the last year the attacks had been much more frequent. They usually occurred about three o'clock in the morning, and lasted for an hour, or an hour and a half.

They appeared to be of a distinctly asthmatic character, and often commenced with a stinging sensation in the throat, and with coughing. The patient had been hoarse for about six months. The impaired respiration was attributed to asthma, and the patient had previously been unsuccessfully treated for that complaint for some months.

A laryngoscopic examination showed two very small growths attached to the edge of the anterior half of the right vocal cord (Fig. 25). The more anterior growth was removed at the second

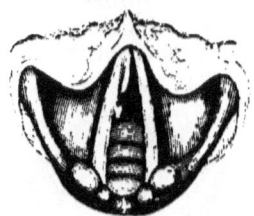

Fig. 25.

visit, but it was not till March, 1866, and after a great number of unsuccessful attempts, that the smaller growth was seized. In May the voice was perfectly clear, and he had had no return of dyspnœa since February 28th.

Dr. Langmaid, of Boston, U.S.A., was present at the removal of the first growth, and this very interesting case was seen by Drs. Tatham, Lanchester, and several other gentlemen, during the time it was under treatment. (The patient applied a year later at the hospital on account of a slight catarrh : he stated that he had had no return of his former symptoms, and the larynx was seen to be quite free from any recurrence of the growths.)

CASE XXII.—*Large Papillary Growth on the Left Vocal Cord;
Partial Evulsion; Great Improvement.*

Eliza W., æt. 45, applied at the Hospital for Diseases of the Throat, January 18th, 1866, on account of complete loss of voice and slight shortness of breath. She stated that for the last twenty-five years she had been hoarse, and that for the last eight her voice had gone altogether ; latterly, she had occasionally suffered from distressing attacks of suffocation. She had been treated by external galvanism, and had been severely blistered on the neck, but without benefit.

On examination with the laryngoscope, a large excrescence was seen occupying the space between the vocal cords. The exact origin could not be made out, but it appeared to grow from the anterior third of the left vocal cord, and occupied the anterior three-fourths of the glottis (Fig. 26).

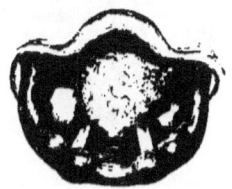

Fig. 26.

The upper opening of the larynx was rather small, and in other respects the case was a difficult one for operative manipulation. Under these circumstances, I was unable to use forceps ; but with a very simple instrument of my own contrivance—a piece of rigid wire bent at a suitable angle, and terminating in a loop—I succeeded in extirpating the principal portion of the growth.

Dr. Andrew Clark, who had made a careful microscopic examination of the specimen, described it as " highly developed, typical, epithelial cancer—pathologically speaking, the most malignant variety he had ever seen, of any *small* growth in that locality." The clinical history of the case did not, however, correspond with the histological deductions : the long time (twenty-five years) that the tumour, or, more correctly speaking, its symptoms, had been in existence, and the total absence of cachexia, indicated a growth of more benign character.

There is a portion of the growth still (Feb. 20th, 1866) in the larynx, but the symptoms are greatly alleviated ; the voice, instead of being completely suppressed, is now only hoarse, and the respiration is not in the least embarrassed.—*Transactions of the Pathological Society*, vol. xvii. page 33.

CASE XXIII.—(*Papillomatous*) *Growths on the Left Vocal Cord and on both Ventricular Bands ; Evulsion of a Large Portion ; no Improvement in Voice.*

Thomas L., æt. 34, coachman, applied at the Hospital for Diseases of the Throat, in January, 1866, on account of aphonia of two years' standing. On laryngoscopic examination, finely-divided excrescences were seen on the left vocal cord and on both ventricular bands, and two nearly smooth, red, mammillary warts on the right vocal cord. The portion of the left vocal cord not covered by growth was considerably congested (Fig. 27, and Plate II. fig. 1).   Many small

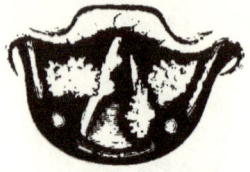

Fig. 27.

fragments were removed with the tube-forceps, and the most anterior wart on the right vocal cord was divided at its origin with the laryngeal lancet.   The patient discontinued attendance after six months, there being but little improvement in the voice.   A portion of the warts still remained both on the right ventricular band and on the left vocal cord.

CASE XXIV.—(*Papillary*) *Growth beneath the Anterior Commissure of Vocal Cords; Treatment by Evulsion; Cure.*

Fanny H., æt. 24, a housemaid, attended as an out-patient at the Hospital for Diseases of the Throat, between the months of February and May, 1866, during which time she was frequently seen by Drs. Tatham and Truell, and Mr. Wilkins.   The patient spoke in a husky voice, and stated that she had been hoarse for eighteen months, that frequently the voice was suddenly lost altogether, and that twice she had had attacks of suffocation.   She also suffered at times from a severe paroxysmal cough.

On laryngoscopic examination, a small pendulous growth was seen beneath the anterior commissure of the vocal cords.   On

Fig. 28.   Fig. 29.

inspiration (Fig. 28) the polypus appeared comparatively small, and quite free of the vocal cords; on phonation it was almost invisible; and on forced expiration (Fig. 29) it rose above the

level of the cords. It was probable, therefore, that the tumour
always prevented the perfect approximation of the cords, thus
causing hoarseness; but that it was only when it was coughed up,
and so became impacted in the glottis, that the aphonia and severe
dyspnœa occurred.

The greater portion of the growth was removed with tube-forceps,
but a small fragment defied all attempts at evulsion (Fig. 30).

Fig. 30.

Under these circumstances, the voice having become quite natural,
and having remained so for three months, the patient did not care
to persevere with treatment.

CASE XXV.—(*Cystic*) *Tumour on the Epiglottis ; Treatment by
Incision and Cauterization ; Cure.*

Maria G., æt. 24, was sent to me by Mr. Gayton, of Brick Lane,
February 20th, 1866. She complained of considerable difficulty in
swallowing during the last six months. Neither the breathing nor
the voice was affected, nor was there any cough. She, however,
experienced a sensation as of a lump in the throat, which obstructed
the food on attempts at deglutition.

The laryngoscope showed an ovoid tumour on the left side of the
upper surface of the epiglottis.

Fig. 31.

On examination with the laryngeal sound, it was found to be
soft, and diagnosed to be cystic. A few small vessels were seen
ramifying on its surface (Fig. 31, and Plate II. fig. 6). The case

18

was watched for some weeks; at the end of June, however, a free incision was made into the cyst, and a quantity of sebaceous-like matter evacuated. A probe coated with nitrate of silver was then introduced for a few seconds. A week later there was no vestige of the tumour, nor of the scar made in incising it.—*Medical Times and Gazette*, 1868, vol. i. p. 631.

CASE XXVI.—*Papillary Growth on the Inter-arytenoid Fold ; Treatment by Evulsion ; Cure.*

Charles B., æt. 35, vocalist, attended at the Hospital for Diseases of the Throat, March 19th, 1866, on account of slight hoarseness in speaking, and complete inability to sing. My attention was called to the case by Dr. Pratt, at that time acting as one of the clinical assistants of the Hospital, who discovered a growth on the fold between the arytenoid cartilages. The patient had attended nine months previously on account of congestion of the larynx, but at that period there were no signs of any growth, and after a short attendance he had resumed his vocation.

On making a laryngoscopic examination, I was able completely to verify Dr. Pratt's diagnosis. There was seen to be a small column-like growth, about the size of a grain of wheat, protruding upwards and forwards from the inter-arytenoid commissure (Fig. 32).

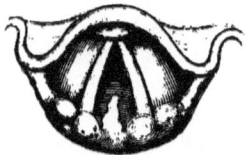

Fig. 32.

The whole of the mucous membrane of the larynx was slightly congested. At the second visit, the growth was easily removed with the tube-forceps.

On microscopical examination, it was found to consist of papillæ covered with squamous epithelium, and of a small amount of connective tissue.

A week later, there was a slight inflammatory thickening of the fold, but no appearance of the growth, and at the end of May the patient was able to accept an engagement at one of the music-halls.

CASE XXVII.—(*Fibrous*) *Growth on the Right Vocal Cord; Treatment by Division of Base of Growth, and Subsequent Removal; Cure.*

Mary R., æt. 57, wife of a stonemason, attended at the Hospital for Diseases of the Throat, March 3rd, 1866, on account of hoarseness, which had existed for three years; occasionally during this period she had completely lost her voice for a day or two.

On laryngoscopic examination, a small, smooth, ovoid growth, about the size of a horse-bean, was seen attached to the centre of the right vocal cord. The growth was situated at the free edge of the cord, and its long diameter was in the antero-posterior direction. The position and connection of the tumour seemed to render it a favourable case for division of the base of the growth. This operation was accordingly performed by means of my laryngeal lancet, at the third visit of the patient, March 14th, in the presence of Dr. Tatham and Dr. Chisholm of Charleston, U. S. There was slight hæmorrhage, and the condition of the larynx could not be clearly ascertained immediately afterwards; but on the following morning the growth was seen to be hanging by a mere shred. It was easily removed by tube-forceps. At the end of a month the voice was quite natural, and the larynx was seen to be perfectly healthy.

CASE XXVIII.—*Fibrous Growth on the Right Ventricular Band; Treatment by Evulsion; Cure.*

Mrs. H., aged 45, was sent to see me at the Hospital for Diseases of the Throat, by Mr. Hind, of Gravesend, April 12th, 1866, with the following history.

For the last four years, she had had a frequent desire to swallow her saliva, and had often experienced great pain in the deglutition of food. During this time the voice had been slightly hoarse. In December, 1865, the breathing had become short, and during the winter she had several times suffered from alarming paroxysms of dyspnœa. Examination with the laryngoscope showed a large, mobile, pedunculated growth, seemingly attached to the right ventricular band. In inspiration, a portion of the right vocal cord could be seen (Fig. 33), but in forced expiration, the growth was pushed upwards and across the larynx, so as not only to conceal the right, but in part also to cover the left vocal cord (Fig. 34). The whole of the tumour was removed in four sittings, with ordinary forceps. This

Fig. 33.   Fig. 34.

case was examined by Dr. Pratt and other gentlemen, both before and after treatment. The larynx is now (June, 1868) perfectly normal, and the patient has since remained quite well.

Dr. Andrew Clark kindly examined this growth, and pronounced it to be of simple fibrous structure.—*Medical Times and Gazette*, 1868, vol. i. page 631.

CASE XXIX.—*Small Papillary Growth on each Vocal Cord; Treatment by Evulsion; Cure.*

Mr. J. P. M., æt. 60, of Lancaster, was brought to me June 29th, 1866, by Mr. Paget.

The patient stated that he had been suffering from gradually increasing hoarseness for three years, and that for the last eight months his voice had become entirely suppressed. His breathing was not at all impaired, nor had he any cough; but he had lately noticed a slighty increased desire to expectorate; his swallowing was not painful, and his general health did not seem at all affected. He stated that four years previously he had suffered severely from pain in the head, the result, he thought, of over-work. He was now, however, gaining flesh, and had a good appetite.

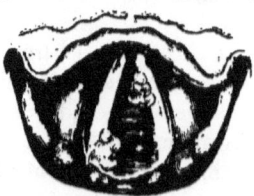

Fig. 35.

Examination with the laryngoscope showed a small round papilliform growth on each vocal cord; the wart on the left cord was about a quarter of an inch from its anterior insertion, and that on the right cord was situated on its cartilaginous portion (Fig. 35).

The small size of the growths rendered operative procedures very difficult, and it took forty sittings to remove the whole of them. Tube-forceps were employed in this case. I have frequently seen this patient since. He has had no recurrence of the growth, but occasionally suffers from congestion of the larynx. He speaks now in his natural voice.—*Medical Times and Gazette*, 1868, vol. i. page 632.

Case XXX.—*(Papillary) Growths on both Ventricular Bands; Treatment by Evulsion; Cure.*

W. S., æt. 49, stableman, applied at the Hospital for Diseases of the Throat in July, 1866, complaining of hoarseness, unaccompanied by any other symptom. My attention was called to the case by my colleague, Mr. Evans, who first saw the patient, and transferred him to me.

On laryngoscopic examination, small pink fimbriated growths were seen projecting from both ventricular bands (Fig. 36). The

Fig. 36.

growth on the right side was removed at the first sitting; that on the left band, after a few unsuccessful and partially successful attempts, on September 10th.

The patient entirely regained his voice, and a fortnight after the last operation there was not the least abnormal appearance in the larynx.

Case XXXI.—*Pedunculated (Papillary) Growth on the Left Vocal Cord; Treatment by Incision of the Base of the Growth; Cure.*

J. N., æt. 23, a street singer, was first seen at the Hospital for Diseases of the Throat in November, 1866. The prominent symptoms, in this case, were dysphonia and a frequent hacking cough, unaccompanied by expectoration. With the aid of the laryngoscope, a small warty growth was seen to be attached by a distinct pedicle to the centre of the left vocal cord.

In the presence of Drs. Taylor, Merryweather, and other gentlemen, the growth was incised with my laryngeal lancet. A week later the growth was not a third of its previous size, and had a grey sloughy appearance. On January 30th, 1867, the larynx was perfectly free from any sign of the growth, though there was still slight congestion. About six months later, that is, on July 18th, 1867, this patient applied again, on account of an attack of quinsey. On recovery from the tonsillitis, his larynx was examined, and there was not the least appearance of the growth : he stated that " he had been able to pursue his profession as an *Ethiopian Serenader* without the least discomfort " since he left the hospital.

CASE XXXII.—*Small (Papillary) Growth on the Left Capitulum Santorini; Treatment by Evulsion ; Cure.*

Catherine H., æt. 50, laundress, applied as an out-door patient at the Hospital for Diseases of the Throat in the latter part of December, 1866, on account of shortness of breath, which was very much increased on exertion. The patient spoke in a husky voice ; but she said that her voice had been so long rather gruff that she scarcely noticed it, and that it was of no consequence to her. She had frequently suffered from sore throat, but the dyspnœa had only come on during the last five months. On laryngoscopic examination, a small pedunculated growth was seen to be attached to the mucous membrane over the capitulum Santorini (Fig. 37).

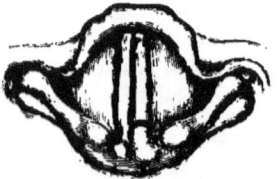

Fig. 37.

This was easily seized and removed with tube-forceps at the second visit, in the presence of Mr. Evans. The patient did not apply again till 1st February, 1867, when all symptoms had passed away, and with them all trace of the growth.

CASE XXXIII.—*Small (Papillary) Growth on the Left Vocal Cord;
Treatment by Evulsion; Cure.*

Mr. S. C., æt. 37, a vocalist, consulted me in December, 1866,
on account of hoarseness, which had prevented him taking any
engagement for the last two years. He had spent two months of
the previous summer at Ems, and had come to me from Torquay,
where he had been recommended to pass the winter, in the belief
that he was suffering from laryngeal phthisis. The patient was
exceedingly nervous and dispirited, and had lost flesh. An additional cause of anxiety existed in the fact that an elder brother, not
long previously, had died from disease of the lungs.

On laryngoscopic examination, a well-defined growth, about the
size of a barleycorn, was seen just in front of the vocal process of
the left vocal cord (Fig. 38), preventing proper approximation in

Fig. 38.

attempted phonation. A careful stethoscopic examination showed
no signs of chest disease; and the patient stated that an opinion
to this effect had been given by several physicians, though no
explanation had hitherto been afforded as to the nature of his
disease. I had the opportunity of showing this case to my colleague, Mr. George Evans, who examined the patient both before
and after treatment.

Great difficulty was experienced in operating in this case, on
account of the patient's nervousness; and it was only after Mr. C.
had sucked ice for a quarter of an hour before each examination,
that the irritability was sufficiently relieved to enable me to introduce
any instrument. Very many unsuccessful attempts, with every variety
of forceps, were made before the growth was removed. Nevertheless, on March 10th, 1867, I succeeded in removing it with
tube-forceps. I have frequently seen this gentleman since. He
has quite recovered his voice, and is able to take engagements, and
to teach singing to a large number of pupils, without the slightest
discomfort or disability.

CASE XXXIV.—*(Papillary) Growth on the Left Vocal Cord;*
Treatment by Evulsion; Cure.

Mrs. C. E., æt. 46, was sent to me by the late Dr. Brinton,
December 14th, 1866, on account of loss of voice, which she had
experienced for six years, with gradually increasing hoarseness for
five years previously. In 1862 she had been examined by Professor
Czermak, but the patient did not know what his opinion had been.

On examination with the laryngoscope, a small pedunculated
growth, about the size of a pea, was seen attached to the anterior
extremity of the left vocal cord (Fig. 39). On the third attempt, the

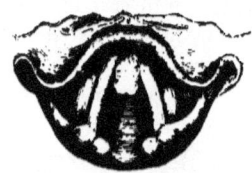

Fig. 39.

whole of the growth was removed with tube-forceps. In May, 1867,
the patient's voice was perfectly natural, and the larynx looked
quite healthy. I had the opportunity of showing this case, both
before and after treatment, to two experienced laryngoscopists,
Dr. Hun, of Albany, and Dr. Lockwood, of New York.—*Medical
Times and Gazette*, 1868, vol. i. page 632.

CASE XXXV.—*Large (Papillomatous) Growth on the Right Vocal
Cord; Treatment by Evulsion; Cure.*

Henry Y., æt. 40, railway porter, from Surbiton, applied at the
Hospital for Diseases of the Throat in the early part of January,
1867, on account of hoarseness. He stated that he first observed
his voice to fail in September, 1865. One of his duties was to call
out the names of the station on arrival of trains. His hoarseness
increased so much that in March, 1866, he was relieved from that
portion of his duty. The voice, however, instead of improving,
gradually became worse. The patient complained of no other
symptom but the hoarseness.

On laryngoscopic examination, a growth, about as large as a
Barcelona nut, was at once seen springing from the middle of the

right vocal cord (Fig. 40). The mucous membrane of the larynx was generally congested. The growth, being in a favourable position,

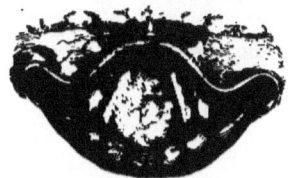

Fig. 40.

was easily removed in a few visits by tube and common antero-posterior forceps.

During the progress of treatment this patient was seen by Dr. Pogojeff, of Odessa, Mr. Du Pasquier, and several other gentlemen at that time attending the practice of the hospital.

CASE XXXVI.—(*Fibrous*) *Growths on the Right Vocal Cord; Excision of small Pieces: no Improvement; Tracheotomy.*

J. K., æt. 54, a gardener, was admitted as an out-patient at the Hospital for Diseases of the Throat, January 17th, 1867, on account of loss of voice and difficulty of breathing. He stated that three years previously he had caught a severe cold, which had affected his throat and chest. Since then he had never quite recovered his voice, and for the last twelve months it had become entirely suppressed. For the past eight months he had suffered from shortness of breath, on the least exertion, so that he had been quite unable to follow his occupation, and latterly he had been several times awakened from his sleep by alarming attacks of suffocation.

On examination with the laryngoscope, the whole of the windpipe was seen to be inflamed and swollen; along the right vocal cord were several round nodulated growths, intimately associated with the structure of the cord. After some weeks of local treatment, in the form of inhalations and topical application of mineral astrin-gents, the inflammation was very much reduced, and attempts at evulsion of the growth were made. The neoplasms were, however, so dense in structure, and so firmly adherent, that the use of forceps proved quite unavailing. Several small pieces were divided with the laryngeal lancet; but this did not give much relief, and the dyspnœa and attacks of strangulation increasing, the patient was admitted into the wards in April. He was unwilling to submit to any further treatment for the removal of the growths; so on April 7th, Mr. Evans performed tracheotomy, and he left the hospital, wearing his tube, at the end of three weeks. This patient presented himself at the hospital in June, 1868, desiring that the canula might be removed; but as the larynx presented a similar appearance to that seen before the operation, this step was not recommended.

CASE XXXVII.—*Papillomatous Growth on the Right Vocal Cord; a smaller one on Under Surface of Epiglottis; Partial Evulsion; Improvement of Voice; Return of Symptoms; Similar Treatment, and Complete Recovery.*

Mr. W. H. H., æt. 40, an overseer of mines, was brought to me in February, 1867, by an able laryngoscopist, Dr. Griffiths, of Swansea. The only symptom was complete suppression of voice. For the last three years he had been quite unable to sound his voice, but previously to that time he had been hoarse.

Examination with the laryngoscope discovered a large highly-divided growth, occupying half of the right vocal cord, and a small smooth wart on the left side of the under surface of the epiglottis (Fig. 41). The latter was immediately removed with the tube-

Fig. 41.

forceps, and the greater part of the growth on the right vocal cord, with the same instrument at a subsequent visit. The voice improved greatly immediately after this operation, and the patient left town a few days later, satisfied with the great improvement that had taken place, but with a very small portion of the growth still attached to the anterior part of the right vocal cord.

May 25th, 1868.—Mr. W. H. H. called on me with a larger growth on the epiglottis than the one I had previously removed, and a recurrence of the excrescence on the vocal cord. These growths were again removed with tube-forceps, and the patient left town in a week with a good voice. A portion of this growth was examined by Dr. Fenwick, who reported that it was of warty character.

In a note received from Dr. Griffiths, December 15th, 1870, he states: "I found Mr. W. H. H. at last. His general health is excellent, and he tells me that his voice is 'capital.'"

CASE XXXVIII.— *Fringe-like (Papillary) Growth on the Posterior Wall of the Larynx; Partial Destruction by Galvanic Cautery; Negative Result.*

Mrs. W., æt. 56, was recommended to consult me in March, 1867, by Dr. Addington Symonds, of Clifton, on account of chronic dysphonia. She complained also of cough, and a tendency to sickness. The symptoms had commenced with a slight attack of laryngitis, accompanied with some dyspnœa; but though, on recovery from the attack, the breathing had become natural, her voice continued hoarse.

On laryngoscopic examination, a pale fringe-like growth was seen projecting from the posterior wall of the larynx, and extending from the level of one cartilage of Wrisberg to the other (Fig. 42, and

Fig. 42.

Plate II. fig. 5). From the easy situation and comparative flatness of the growth, it seemed a most favourable case for electric cautery. The pointed laryngeal instrument, with a four-celled Smee's battery, was used. Two days later, it was found that about a third of the growth on the right side was destroyed, and that in its place there was a greyish slough; the patient, however, complained of having experienced such severe pain in the ears after the first cauterization, that she would not consent to any further treatment of a similar kind.

I saw her several times afterwards, but there was still thickening of the posterior wall of the larynx, and she did not appear to have derived much benefit from the treatment.

Case XXXIX.—*Small Fibroma on the Anterior Portion of the Left Vocal Cord; Slight Alteration in Voice; Treatment by Evulsion; Cure.*

Mr. T., æt. 27, a vocalist and comedian, from New York, was recommended through Dr. Marion Sims to consult me on his arrival in England. He called on me March 15th, 1867, and stated that his singing voice had been entirely lost for two years, though previously he was considered a powerful tenor. His ordinary speaking voice was not apparently affected, though he fancied it had changed tone, and frequently it required a considerable effort on his part to make himself heard on the stage.

On laryngoscopic examination, a small, smooth, bright red, polypoid, growth was seen on the upper surface of the left vocal cord, close to the anterior commissure (Fig. 43). The neoplasm did not

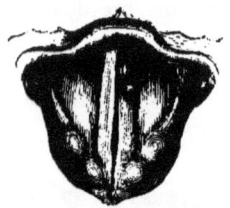

Fig. 43.

project beyond the free edge of the vocal cord, but appeared to rotate slightly from above downwards. The vocal cords themselves were quite normal. The movement of the growth led to the supposition that it had a narrow peduncle attached either to the vocal cord or to the lower edge of the ventricle. At the second visit, the growth was seized with the tube-forceps, and easily removed. Examined microscopically, the growth was found to consist entirely of white fibrous tissue.

I saw Mr. T. six months later, and he had then quite regained his singing voice, though he had lost some of his higher notes.

Case XL.—*Symmetrical (Papillomatous) Growths on both Vocal Cords; almost Complete Evulsion; Great Improvement.*

Henry S., a labourer, æt. 32, applied at the Hospital for Diseases of the Throat, March 21st, 1867, on account of complete loss of voice. He had been hoarse for nine years, and for two years and a

half the voice had been altogether suppressed. His breathing was
not at all impaired.

The laryngoscope revealed a cauliflower-looking growth on each
vocal cord. The neoplasms, which were symmetrical, were situated
at the junction of the anterior third with the posterior two-thirds of
the cords (Fig. 44). The patient denied ever having had syphilis.

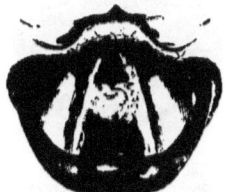

Fig. 44.

On the first visit, the whole of the growth on the right vocal
cord was removed with the common antero-posterior forceps, in the
presence of Dr. Padley, of Swansea. On the next visit, March 28th,
there was no perceptible improvement in the voice. On examining
the larynx, the mucous membrane was seen to be in a state of
extreme congestion ; and it was deemed advisable to defer any
operation on the remaining growth. It was not till April 25th that
I was able to introduce an instrument within the larynx, when, in the
presence of Dr. M'Call Anderson, I removed what appeared to be
the whole of the growth on the left cord. Three days later, however, a very small portion was still seen to remain. The patient did
not attend again for three weeks, when he presented himself as
cured. The voice, though strong and phonetic, was not quite
clear, and I could see that in phonation the cords did not
accurately approximate, on account of the small piece (not larger
than a pin's head) of the growth that remained on the left side.
The patient, however, was so satisfied with his condition, that he
declined to have any further treatment.

CASE XLI.—*Fibro-cellular Polypus on Under Surface of the Epiglottis ; Treatment by Galvanic Cautery ; Cure.*

Harriett T., æt. 28, servant, applied at the Hospital for Diseases
of the Throat in the latter part of March, 1867, on account of
hoarseness, which had been coming on since May, 1865, and had
been attributed by her medical attendant to bronchitis.

On laryngoscopic examination, the cause of the dysphonia was at once apparent, for a pale, semi-transparent pedunculated tumour was seen growing from the right side of the under surface of the epiglottis, near to its free edge (Fig. 45).

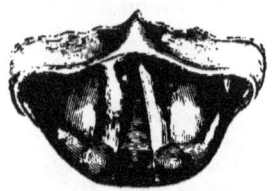

Fig. 45.

As at that time I was making experiments with the galvanic cautery, I determined to employ this agent.

Before operating, I demonstrated the case for several successive Thursdays to the various medical practitioners at that time attending the Hospital. The patient was seen by, amongst others, Dr. Atkinson and Dr. Maccaldin. On May 14th I passed an electric-cautery wire, connected with a Middeldorpf battery, round the growth, and on application of the current, it at once dropped off, and was expelled into the pharynx. On examination after removal, the growth was seen to be perfectly white, and contained a gelatinous substance like that found in nasal polypi. Four hours after the operation, the patient applied to me, on account of slight pain and difficulty in swallowing.

On making a laryngoscopic examination, the orifice of the larynx was seen to be rather congested, and the edge of the epiglottis slightly œdematous. By sucking and swallowing ice for a few hours, the symptoms were relieved, and two days afterwards, the operation was found to have been perfectly successful. A month later, even the situation of the growth could not be discovered.

CASE XLII.—(*Fibroma*) *on Superior Surface of the Right Vocal Cord ; Unsuccessful Attempts at Evulsion ; Treatment by Galvanic Cautery ; Subsequent Laryngitis ; Ultimate Cure.*

The Rev. C. F. D., a missionary, who had been many years in the Bombay Presidency, but had been obliged to suspend his labours during the last two years, on account of gradually failing voice, applied to me, May 5th, 1867. He had been recommended to try change of climate to Europe, and had had his uvula removed. No laryngoscopic examination had, however, been made up to the time he came to me. On using the mirror, a smooth and but slightly elevated growth was seen covering the middle third of the right vocal cord. In vocalization it impinged against the left cord, so that neither could completely advance to the median line (Fig. 46, and Plate II. fig. 4).

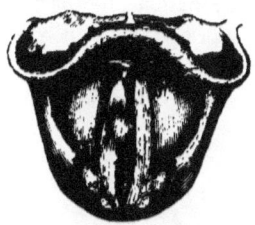

Fig. 46.

Several unsuccessful attempts at removal having been made with a great variety of growth-forceps, it was determined to employ the galvanic cautery. This was accordingly done on July 14th. My assistant, Mr. Lennox Browne, attended to the battery, whilst I employed the pointed electrode. The same evening the patient suffered from great dyspnœa, and had symptoms of acute laryngitis : the whole of the mucous membrane of the larynx was seen to be greatly inflamed, and the right ventricular band was œdematous. I was anxious to scarify the part, but the patient would not consent to it. He was directed to use warm inhalations of compound tincture of benzoin, and the next morning I found him considerably relieved. There was still so much congestion of the larynx, that the condition of the right vocal cord could not be seen. On the 16th a black eschar was observed to occupy the greater part of the vocal cord which had been touched. A week later, the cord had a uniform grey appearance. The patient was not seen again till the 20th of September, and at that time, with the exception of slight redness of the right vocal cord, the whole of the larynx was perfectly healthy.

CASE XLIII.—*Papillary Growth on the Posterior Wall of the Larynx; Treatment by Evulsion; Cure.*

Eliza C., æt. 53, married, applied at the Hospital for Diseases of the Throat, May 10th, 1867, on account of shortness of breath and loss of voice. She stated that she had frequently suffered from quinsey, but that difficulty of breathing had first come on eighteen months previously. Her voice first became hoarse a year ago, and had only been altogether suppressed during the last seven months. She also complained of a frequent cough, which was apparently caused by an itching sensation in the throat. There was no evidence of the existence of either phthisis or syphilis.

On laryngoscopic examination, the larynx was naturally very small, and the posterior half of the glottis was seen to be occupied by a white cauliflower excrescence growing from the posterior wall of the larynx (Fig. 47). The lungs were examined, and appeared to be healthy.

Fig. 47.

The case being urgent, treatment was at once adopted, and a large fragment was removed at the first visit with tube-forceps, in the presence of Dr. Longmarsh and Staff Assistant-Surgeon Semple. A week later, the breathing was found to be much improved, and only a small portion of the growth remained. This was easily removed with the same instrument.

The growth was examined microscopically by Dr. Andrew Clark, and pronounced to be "a simple papillary formation."

CASE XLIV.—*Papillary Growth on the Right Ventricular Band ; Treatment by Evulsion ; Cure.*

Sarah F., æt. 41, a married woman, came under my care at the London Hospital, in June, 1867, on account of hoarseness of six years' standing. The voice was not completely lost, but the patient spoke very gruffly. She also complained of occasional attacks of dyspnœa, and stated that she had twice lately awoke early in the night with a feeling of suffocation. She had a slight tickling cough, unaccompanied with expectoration. On laryngoscopic examination, a cauliflower-like growth, about the size of a bean, was seen on the right ventricular band. On inspiration the growth covered the posterior four-fifths of the right vocal cord (Fig. 48), but on attempted vocalization it projected across the glottis, and covered the middle third of the left vocal cord (Fig. 49). Two unsuccessful

 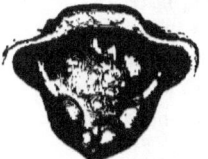

Fig. 48.   Fig. 49.

attempts were made to seize the growth in July, but in the following September the whole of it was removed with tube-forceps. The tumour was examined microscopically by Dr. Andrew Clark, and pronounced to be a simple papilloma.

CASE XLV.—*Large Papillomatous Growths on both Vocal Cords; Evulsion of a Portion; Discontinuance of Treatment on account of Pregnancy of Patient; Subsequent Tracheotomy.*

Mrs. J. M. B., æt. 30, a watchmaker's wife, from Clapton, was admitted into the Hospital for Diseases of the Throat, June 13th, 1867, on account of great dyspnœa, accompanied with aphonia. The aphonia had existed three years, but the dyspnœa only a few weeks. The patient was pregnant. On laryngoscopic examination, both vocal cords were seen to be covered with large cauliflower excrescences. The mass on the right vocal cord grew upwards and completely occupied the right half of the cavity of the larynx (Fig. 50). I removed several large pieces with tube-forceps, but

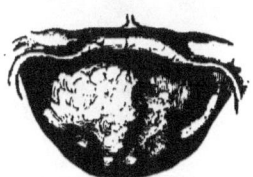

Fig. 50.

the patient was obliged to leave the hospital in the early part of September, her general condition at the time rendering further treatment difficult. On the 24th of November her respiration was so seriously embarrassed, that Mr. De Berdt Hovell, of Clapton, performed tracheotomy. She recovered well, and was delivered of a still-born child, January 4th, 1868.

Mr. Hovell kindly communicated to me (April 5th, 1870) the following notes of her condition :—

" Her present condition is pretty well. The tube is still in ; she has recovered her voice, although not its full power ; she has been confined, as you know, of a dead child since the operation, and is expecting her confinement again next month."

CASE XLVI.—*Numerous (Papillomatous) Growths in the Larynx; Treatment by Evulsion ; Recovery of Voice ; Subsequent Recurrence ; Cure.*

Miss M., æt. 30, the sister of a physician, consulted me, May 2nd, 1867, on account of loss of voice and shortness of breath. She stated that eleven years ago she first became hoarse. Previous to that time she had a soprano voice, and was accustomed to sing a good deal. She thus described the development of the hoarseness :—
" My voice altered, first losing high notes, then becoming a whisper, and wavering from time to time at every change of weather and climate." For the last two years she had occasionally noticed that her respiration was embarrassed, and she complained of a frequent desire to clear the throat. On making a laryngoscopic examination, the whole of the interior of the larynx, with the exception of the epiglottis, was seen to be covered with small warty excrescences, and one small growth was seen to be attached beneath the anterior commissure of the vocal cords (Fig. 51). Owing to the small

Fig. 51.

size of the larynx, the treatment was carried out with some difficulty, with the aid of tube-forceps, and it required two months, with almost daily sittings, to clear the larynx.

The patient left town with a good voice, and with perfectly easy respiration.

This lady called on me, June 27th, 1870, and I found there was a recurrence of the growth on the ventricular band and vocal cord, on the right side. Her voice was phonetic, but rather shrill. Treatment was again pursued with tube-forceps, and early in March, 1871, the larynx was again completely cleared.

Case XLVII.—(*Papillary*) *Growths on the Vocal Cords ; Treatment by Evulsion ; Cure.*

John T., æt. 45, a shoemaker, applied at the Hospital for Diseases of the Throat, July 9th, 1867, on account of a bad cough, accompanied by hoarseness. He stated that for some years he had suffered from an irritable cough, with a desire to clear the throat of some obstruction that seemed to be constantly present. For the last eighteen months the cough had increased in frequency and severity, and the attacks sometimes lasted for many minutes. The cough was always excited when he lay down, or if he attempted to stoop over his work. The hoarseness had only commenced about a year and a half previously ; it had gradually increased, until, at the time of application at the Hospital, his voice was entirely suppressed.

On laryngoscopic examination, a large warty growth was seen projecting from the middle third of the right vocal cord, and on the left side there was another irregular neoplasm growing from the edge of the cord. The cords were generally jagged and irregular in outline, and the mucous membrane of both pharynx and larynx was much congested and relaxed. It was not till after some weeks of local treatment by mineral astringents that the congestion of the larynx was sufficiently reduced to admit of attempts at removal of the growth. The tumour on the right side was eradicated with the common antero-posterior forceps at the first attempt ; but the growth on the left side required sixteen sittings, at which both the common and tube-forceps were used, before it could be entirely removed. After evulsion of the larger portions, the cough quite ceased, but it was not till the larynx had been quite clear for some months that the voice was completely restored.

This case was seen during treatment by Surgeon-Major Trestrail and the Rev. David Bell, M.D., of Goole, as well as by many other practitioners.

CASE XLVIII.—*Fibro-cellular Growth on the Under Surface of Epiglottis ; Treatment by Evulsion ; Cure.*

Joseph L., æt. 30, a furniture-dealer, applied at the Hospital for Diseases of the Throat, July 18th, 1867, on account of loss of voice, of nine months' duration. He served in an open shop, and was a good deal exposed to cold and draughts. On making a laryngoscopic examination, a small, red, irregular, pedunculated, growth was seen attached to the under surface of the epiglottis, close to the commissure of the vocal cords (Fig. 52). On the second visit

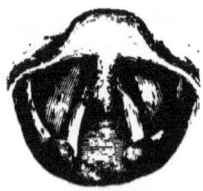

Fig. 52.

(July 21st), in the presence of Surgeon-Major Trestrail, I removed the whole of the growth with tube-forceps. A week later (July 28th) the voice was entirely restored, and there was no appearance of the growth. On microscopic examination, this growth was found to consist of fibro-cellular tissue, in which, on the addition of acetic acid, numerous nuclei were apparent. There was a small amount of interstitial fluid, and in it were a few nucleated cells.

CASE XLIX.—*Fasciculated Sarcoma on the Under Surface of the Epiglottis ; Treatment by Galvano-cautery and Evulsion ; Great Improvement ; Recurrence.*

Anne H., æt. 53, unmarried, a Bible-reader, first came under my notice in July, 1869, complaining of loss of voice and extreme dyspnœa. She experienced great pain in attempting to speak, and there was also so much pain in swallowing, that for some time she had only been able to take food of fluid, or semi-fluid consistence. On laryngoscopic examination, a smooth, red, neoplasm, about the size of a kidney-bean, was seen on the under surface of the epiglottis (Fig. 53, and Plate III. fig. 3). The position of the growth was very similar to that in the last case ; but in this instance it was much larger and had a broad base. On making an examination with the laryngeal sound, the polypus was found to be unusually hard. Several attempts were made to remove it with forceps, but, owing to the gristle-like consistence of the

Fig. 53.

tumour, they only resulted in the tearing away of small portions of superficial mucous membrane. It was therefore determined to use the electric cautery. The first application was made on January 16th, 1868, with the aid of my assistant, Mr. Lennox Browne. The patient experienced but little inconvenience from this treatment, and, ten days afterwards, the growth seemed to be sensibly diminished in size. A second application was made March 5th, and though some pain was felt after the operation, it appeared to have a decidedly beneficial effect on the size of the growth. In the summer of 1868 the treatment had been so successful, that the patient was able to take ordinary food, and her voice was almost restored to a natural tone, so that she was able to resume her occupation in the London Female Bible Mission. This entails not only a good deal of talking and reading, but also exposure to all kinds of atmosphere and temperature. In January, 1870, a recurrence of the growth having taken place in the same situation, I removed a piece with my cutting forceps; it was exceedingly hard and difficult to tear away. On microscopic examination by my brother, Mr. Stephen Mackenzie, it was seen to consist, in parts, of simple fibrous tissue; but in others, abundance of nuclei were found, amongst long interlacing fibres (Plate I. fig. 7). This growth resembled the fibro-nucleated tumour of Dr. Hughes Bennett.

There is still a slight projection from the under surface of the epiglottis, but it is scarcely sufficiently defined to warrant any further interference.

CASE I.—(*Papillary*) *Growth on the Right Vocal Cord ; Treatment by Evulsion ; Cure.*

Mr. S., æt. 64, a merchant, was recommended to consult me by Mr. Ince, September 10th, 1867, on account of aphonia, of three years' standing. He had, however, been hoarse, to a slight extent, for four years, and during that time had tried a great variety of medical and climatic treatment. He attributed his affection to a severe cold, which he had caught whilst travelling by railway at night.

On laryngoscopic examination, a small mammillated growth of dark red colour was seen springing from the right vocal cord, at about its middle third.

The greater portion was removed on the first attempt, but a small fragment, scarcely larger than a pin's head, remained. The voice did not at all improve until January, 1868, when, after a number of unsuccessful attempts, I succeeded in removing the remaining portion, my assistant, Mr. Lennox Browne, being present at the operation.

CASE LI.—(*Papillary*) *Growth on the Right Vocal Cord ; Evulsion and Crushing ; Cure.*

W. W., æt. 55, a druggist's dispenser, from Greenwich, applied at the Hospital for Diseases of the Throat, in October, 1867, on account of complete loss of voice, which had come on quite suddenly three years before. For the last ten years he had occasionally lost his voice for a day or two, but had always recovered it. He had also lately, on two occasions, awoke suddenly in the night with a sensation of suffocation. He had suffered from constitutional syphilis, from time to time, since he was twenty-six years old.

On laryngoscopic examination, a smooth white growth, the size of a haricot bean, was seen attached to the right vocal cord, which was considerably congested. The vocal cords remained widely apart on attempted phonation.

A considerable fragment was removed with common antero-posterior forceps, in the presence of Surgeon-Major Trestrail and other practitioners, on October 24th, but the remaining portion was of so dense a structure, and so firmly adherent, that it could not be removed, except with more force than was considered justifiable. After repeated crushings, however, the growth gradually atrophied, and at Christmas there remained only a slight unevenness of the right vocal cord. The cords approximated well, but were slightly congested ; the voice was fairly good.

CASE LII.—*Fibro-cellular Growth on the Right Vocal Cord; Treatment by Evulsion; Complete Removal at first Operation; Cure.*

Mrs. R., æt. 65, consulted me on the 8th of October, 1867, on account of loss of voice and difficulty of breathing, which had been coming on since the year 1852. For the last seven years she had been very hoarse, but during the last twelve months her voice had been entirely suppressed.

On making a laryngoscopic examination, a large, red, globular growth was seen occupying the anterior three-fourths of the glottis (Fig. 54, and Plate II. fig. 10). It was very mobile, and therefore

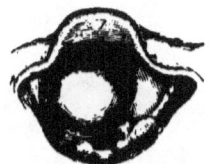

Fig. 54.

judged to be pedunculated ; but as its base was entirely hidden, its origin could not be ascertained. On the 30th of October, the entire growth was removed with the common laryngeal forceps. Immediately after its removal, the patient spoke in a clear natural voice, but she became a little hoarse the same evening. The next day it was seen that the growth had been attached to the posterior part of the right vocal cord. A slight roughness and hyperæmia in this situation indicated its previous base. Ten days later, the larynx was seen to be perfectly healthy, and the voice and breathing natural.

The growth (Plate II. fig. 10) was at that time the largest I had removed through the mouth, without previously performing tracheotomy. It measured, on removal, more than six-eighths of an inch in its long diameter, and one inch and three-eighths in circumference. Pathologically it was considered a fibro-mucous polypus.—*Transactions of the Pathological Society,* vol. xxi., January 21st, 1868.

CASE LIII.—*(Papillary) Growths along the whole extent of both Vocal Cords; Partial Evulsion; Improvement.*

W. H., æt. 26, a journeyman butcher, applied at the Hospital for Diseases of the Throat, November 6th, 1867. Five years previously he had contracted syphilis, and ten weeks after the primary affection, had had sore throat and a copper-coloured eruption on the skin. Two years ago he had a bad ulcerated sore throat, and his voice became very hoarse. Nevertheless he had continued his occupation (which involved the constant use of his voice outside a shop in the evening) until two months previous to his coming to the Hospital.

On laryngoscopic examination, the mucous membrane of the larynx was seen to be in a state of extreme hyperæmia, and both vocal cords were covered with small irregular growths, of a bright-red colour (Fig. 55). Owing to the patient being very nervous,

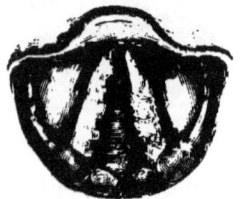

Fig. 55.

and unable to remain still, several unsuccessful attempts were made before any portion could be removed, and he was consequently very much discouraged. It was not till February 6th that two or three small pieces were removed from the right vocal cord, in the presence of Surgeon-Major Trestrail. On the 13th another piece was removed. On the 20th his voice was found to be somewhat improved. The right vocal cord was then quite clear, but the posterior half of the left vocal cord was still covered by an irregular excrescence. The patient did not attend again.

CASE LIV.—(*Benign Epithelial*) *Growth on the Right Vocal Cord ; Treatment by Caustic and Astringent Solutions ; Cure.*

James B., æt. 43, a small farmer, applied at the Hospital for Diseases of the Throat, December 12th, 1867, on account of hoarseness, which had existed for fourteen months. On making a laryngoscopic examination, a pale and very smooth growth, about the size of a split tare, was seen projecting from the free edge of the right vocal cord, at its posterior third. The patient was kept under observation for two months, without any treatment, in order that the progress of the case might be watched ; but the tumour did not undergo any change. Strong solutions of nitrate of silver, iron-alum, and persulphate of iron, were then carefully applied to the tumour about twice a week for six months. At the end of this time the tumour had gradually disappeared, and the voice, which had been slowly improving, had become quite clear. I may mention that in this case the small swelling remained visible, when the vocal cords were in a state of tension, for several weeks after it had ceased to be seen when the cords were relaxed. This case being a rather remarkable one, was repeatedly brought under the notice of the medical practitioners who, during the patient's course of treatment, were attending the hospital. It was especially known to Surgeon-Major Trestrail and Mr. Lennox Browne.

CASE LV.—*Pedunculated (Papillary) Growth on the Left Vocal Cord; Treatment by Evulsion ; Cure.*

W. K., æt. 13, a washerwoman's son, applied at the Hospital for Diseases of the Throat, December 16th, 1867, on account of shortness of breath and hoarseness.

The mother stated that her boy had an attack of croup when three years old, and since that time he had frequently suffered from sore throat. In 1865 she had brought him to this hospital, and at that time he had been treated for inflammation of the windpipe. Six months previous to his present application he had had scarlet fever, which was followed by enlargement of the glands of the neck. She stated further, that, occasionally, he had a croupy cough at night, and that his breathing was often embarrassed. The boy complained of a feeling of dryness, and, at times, of a sensation of choking. The voice was gruff, and, at intervals, altogether suppressed. On laryn-

21

goscopic examination, a wart-like growth about the size of a pea was discovered projecting from the middle of the left vocal cord.

No attempt at removal was made at the first visit of the patient, but a week later, the growth was easily seized with tube-forceps, and the whole of it removed, in the presence of Drs. Welch and Peléchin.

He was seen at the end of March; the larynx was clear, his voice normal, and he was perfectly well in every respect.

CASE LVI.—*Fibroma on Under Surface of Epiglottis; Treatment by Evulsion; Cure.*

Henry W., æt. 34, a sugar-baker, applied at the London Hospital in December, 1867, on account of difficulty of swallowing and a constant disposition to clear his throat. These symptoms had been coming on for two years and a half, but had become much worse during the last ten months. A laryngoscopic examination showed a smooth, pale red, growth on the left side of the larynx. The origin of the neoplasm, which was about the size of a haricot bean, could not be made out; but it was believed, at the time, to be attached to the epiglottis. On the 4th of December the entire growth was removed with common laryngeal forceps. The case was seen before and after treatment by Dr. Alexander Fox. Examined microscopically, the growth was found to be of an obscurely fibrous structure.

CASE LVII.—*(Papillary) Growth on the Left Vocal Cord and on Epiglottis; Treatment by Evulsion and Crushing; Cure.*

W. W. R., æt. 17, an engine-fitter, from Barnstaple, was sent to me by Dr. Johnston, of that town, in December, 1867, on account of loss of voice, which had existed for two years. The boy was a skilled player on the cornet. He complained of dryness, and a sense of obstruction in the throat; he had a slight cough, and his voice was completely lost. There was no dyspnœa.

On laryngoscopic examination, a small, bright red, growth, the size of a tare, was seen on the under surface of the epiglottis, on the right side; and an irregular warty growth, of pink colour, extended nearly the whole length of the left vocal cord (Plate II. fig. 7). This growth was very moveable; in inspiration, it was drawn down, and only occupying half the area of the glottis, looked about the size of a large bean (Plate II. fig. 7); but in forced expiration, the growth occupied nearly the whole of the opening of the glottis, leaving only two minute chinks, one at the anterior and right side, and another at the posterior part of the larynx (Plate II. fig. 8).

The boy was admitted into the Hospital for Diseases of the Throat, and remained there nearly four months, during which time a great number of operations were performed. The small polyp on the under surface of the epiglottis was at once eradicated. The larger growth, however, had a long and membranous base, and yielded to the slightest touch with an instrument, so that, when an attempt at evulsion was made, it rotated completely beneath the vocal cord, and left nothing but its thin membranous attachment visible (Plate II. fig. 9). This constituted a great difficulty, and accounts for the long time the boy remained under treatment. Ultimately, however, the growth was entirely removed, and he left the Hospital with his voice completely restored. Tube-forceps, common forceps, and Stoerk's écraseur were used in this case. Shortly after the boy returned home, I was pleased to receive a letter from Dr. Johnston, acknowledging the satisfactory issue of this very difficult case.

I saw this patient again in March, 1869. There was no return of the growth ; the larynx was normal in every respect, and the voice perfectly natural.—(*Medical Times and Gazette*, vol. i. 1868, page 632.)

CASE LVIII.—*Benign Epithelial Growth on the Right Ventricular Band ; Treatment by Evulsion ; Cure.*

F. S., æt. 10, an artisan's child, was brought to me at the Hospital for Diseases of the Throat, by his father, on January 7th, 1868, on account of loss of voice and shortness of breath, which had existed since the child was six months old. At that age he had an attack of what was pronounced to be croup. Since then he had suffered from occasional severe attacks of suffocation at night, and once the father had been advised to take him to Middlesex Hospital, in order that the windpipe might be opened. The child remained in that institution a few days, but the symptoms subsided, and no operation was recommended.

A laryngoscopic examination was easily made, and a red, fimbriated, neoplasm, about the size of a kidney-bean, was seen on the right ventricular band, near the edge of the ventricle. Several minute pieces were removed at various times, but it was not till February 13th that the last remaining portion was removed with the tube-forceps, in the presence of Dr. Alexander Hewan, Mr. Pugin Thornton, and others. On microscopic examination, the growth was found to consist entirely of epithelial cells in various stages of development. No connective tissue, nor papillæ, could be discovered in the tumour. On March 1st, the voice was completely restored, and there was no trace of the growth.

CASE LIX.—*Fasciculated Sarcoma on the Right Vocal Cord and in the Right Ventricle; Treatment by Evulsion; Constant Recurrence.*

R. P., æt. 42, a carman, was sent to me at the Hospital for Diseases of the Throat, January 13th, 1868, by Dr. Woolley, of Kentish Town, on account of loss of voice.

On laryngoscopic examination, a white nodule, the size of a pea, was seen growing on the right vocal cord, at about its middle. The origin of the growth was merged in the substance of the vocal cord, there being no peduncle, line of demarcation, nor difference in colour. There was also a hard white nodule in the submucous tissue of the tongue, at about its centre. There was an obscurely syphilitic history; and on this account, and also because the growth was unfavourable for removal, the patient was treated for some time by the internal administration of iodide of potassium. The little tumour, however, underwent no improvement. In September, after a few weeks' absence, I saw that the growth had considerably increased in size, and that it had become irregularly lobulated. Several small particles were at different times removed. One of these pieces I sent with the patient to Dr. Fenwick, requesting him to be kind enough to compare it with the growth in the tongue, from which he removed a small slice. He reported to me that both growths consisted of a simple fibrous tissue. On several attempts at removal, the growth was found to be so unyielding that I was content with crushing it. This procedure generally resulted in atrophy of small portions. In the summer of 1869 the tumour had assumed a distinctly pedunculated character, and I removed a piece, about the size of a large pea, in the presence of Mr. Wordsworth. The patient attended from time to time, and small pieces of growth were generally removed at each visit. By this means the development of the growth was checked; but it was not entirely eradicated : it indeed showed a constant inclination to extend. The fresh growth seemed to sprout from the ventricle, and I came to the conclusion that the neoplasm was partly situated in that cavity.

December, 1870.—On the 10th of last month (November) I removed a piece about the size of a Barcelona nut. Half of it was white, with little tentacle-like projections, the other half was quite pink, and apparently composed of closely-packed columns. On the following Thursday, November 17th, I removed another piece quite as large, but black and sloughy in appearance on its free surface. On November 24th a third large fragment was evulsed. This piece was quite soft and friable. After each operation the vocal cord appeared almost cleared of growths; but at the next visit a new formation had taken place.

The later portions removed were examined by Mr. Stephen Mackenzie, and pronounced to be in some parts fibrous, but in others to consist of the elongated, oat-shaped, granular, nucleated cells which characterize fasciculated sarcoma (Plate I. fig. 6). In some parts, also, abundant free nuclei were found, and in others simple epithelial cells. These specimens were also carefully examined by several of my colleagues at the London Hospital.

P.S. *February 9th*, 1871.—I saw this patient after an interval of two months. He informed me that he had lately been in King's College Hospital with dropsy, under the care of Dr. Lionel Beale. There was a slight fresh development of the growth; and in the presence of Dr. Griffiths, of Swansea, and Dr. Roberts, of University College Hospital, I removed a piece the size of a haricot bean.

CASE LX.—*Papillary Growth attached to the Right Vocal Cord; Treatment by Evulsion; Cure.*

M. J., æt. 23, a nursemaid, applied at the Hospital for Diseases of the Throat, February 6th, 1868, on account of shortness of breath, which had existed fourteen months, and was believed to have originated in catarrh.

On laryngoscopic examination, a long filiform growth was seen to be attached to the right vocal cord, near its anterior insertion; the growth occasionally hung down in the glottis (Fig. 57), and at other times became lodged between the vocal cord and the ventricular bands (Fig. 56). At the second visit, February 13th, the growth was easily seized with common laryngeal forceps, and entirely removed, in the presence of Surgeon-Major Trestrail and Dr. Wilkie.

 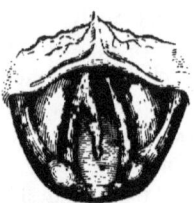

Fig. 56.   Fig. 57.

On microscopic examination, the growth was found to consist almost entirely of squamous epithelium, one very large papilla only being found at the extremity of the growth.

The patient was seen on the 27th, when the larynx was clear, and there had been no recurrence of the dyspnœa.

CASE LXI:—*(Papillary) Growth on the Posterior Part of the Right Vocal Cord; Partial Evulsion; Improvement; Recurrence of Growth; Complete Evulsion; Cure.*

Mrs. H., æt. 31, consulted me (by the advice of Mr. Harston, of Islington), February 8th, 1868, on account of aphonia of a year's duration. An examination with the laryngoscope showed a deeply fissured excrescence, about the size of a small raspberry, on the posterior part of the right vocal cord (Fig. 58). The growth

Fig. 58.

generally completely prevented the approximation of the cords, but sometimes it flapped up, and a feeble phonetic sound could be produced. The throat being exceedingly irritable, and the lady very nervous, it was not till after several months' treatment with tube and common forceps that the greater part of the growth could be removed. A fairly good voice was recovered in June, but a small portion of the growth still remained. It will be readily understood that constant operative treatment on the larynx is apt in nervous people to cause a certain amount of " wear and tear," and at the end of July, Mrs. H. seemed rather out of health. I therefore advised her to discontinue her visits, as may be seen by reference to the *Medical Times and Gazette*, vol. i. p. 632, in which I gave a short account of this case. At the time, I expressed an opinion that a further development of growth was probable; my anticipation proved correct. In the following spring the lady came again under my care, and after considerable trouble, the entire growth was completely removed. In July, 1870, there was no appearance of any recurrence.

CASE LXII.—*Papillary Growths on the Right Vocal Cord; Partial Evulsion; Result unknown.*

Elizabeth F., æt. 57, a fish-hawker, came under my care at the London Hospital, in February, 1868, on account of cough and hoarseness, which had existed for four years. A laryngoscopic examination showed a number of small whitish warty growths, forming a fringe along the entire length of the free edge of the right vocal cord.

Two small fragments were removed at the second visit, but the patient did not apply again. The particles removed were of papillary structure.

CASE LXIII.—(*Papillomatous*) *Growth in the Larynx; Treatment by Evulsion; Cure.*

Master Sydney D., æt. 10, was sent by Mr. Graves, of Gloucester, in February, 1868. His father stated that his voice had been quite natural until six months previously; at that time it became slightly hoarse, and for the last few weeks it had been entirely lost. He also suffered from shortness of breath on exertion, and had lately been awakened from sleep by attacks of suffocation.

Fig. 59.

Examination with the laryngoscope showed a white cauliflower-like growth, occupying almost the entire glottis (Fig. 59). On measurement, it was found to be five-eighths of an inch in length and half an inch broad.

Never having known so large a growth to take place in so short a time, I examined most carefully into the previous history of the patient, and thoroughly satisfied myself that there had been no symptoms of laryngitis, or neoplasm, previous to the hoarseness, which had come on only six months before I saw him.

My little patient was exceedingly nervous, so that at first nothing could be done. Two unsuccessful attempts to remove the growth were made under chloroform, and for a long time after he came under treatment, I had to be content with simply introducing the mirror, and getting a more or less partial view of the larynx. After some months he lost his fear, and the whole of the growth was ultimately removed with tube and common forceps. The voice, however, remained completely suppressed for several weeks after the larynx was clear of growth. This was probably on account of some congestion of the vocal cords. When restored, the voice remained harsh and disagreeable for three or four months, but it ultimately became perfectly clear. I saw this young gentleman in April, 1870, and there was not then either a sign of the growth, or the least abnormality in the strength and tone of the voice.— (*Medical Times and Gazette*, 1868, vol. i. page 632.)

Case LXIV.—(*Papillomatous*) *Growths on the Vocal Cords Partial Evulsion; Tracheotomy; Thyrotomy; Cure.*

I was called down to Norwood on April 21st, 1868, to see Miss B., æt. 66, who, the night previously, had suffered from a severe attack of dyspnœa. On further inquiry, I found that she had lately had frequent attacks of dyspnœa of a very severe character, and that she had lost her voice since the year 1854. She was in good general health, though suffering from slight cardiac weakness.

On laryngoscopic examination, I found a large, bright red, lobulated growth, about the size of a gall-nut, blocking up the anterior two-thirds of the glottis (Fig. 60, and Plate II. fig. 11). The insertion could not be accurately ascertained, but it was believed that the mass consisted, in point of fact, of two growths, springing from the anterior two-thirds of each vocal cord.

By my recommendation, Miss B. came to town on April 29th. On six occasions after this date, I removed large fragments of growth with forceps, through the mouth; but after operating on her on May 18th, so much inflammatory swelling took place, that I did not consider it safe to pursue this method of treatment any further.

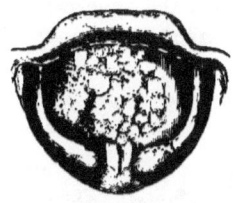

Fig. 60.

The respiration being still embarrassed, I determined to have tracheotomy performed, and the remainder of the growth removed by thyrotomy.

On the evening of May 21st she had another very severe attack of dyspnœa. The attacks came on whenever she began to doze, so that she could obtain but very little refreshing sleep.

The next day, May 22nd, the symptoms being no better, I requested my colleague, Mr. Couper, to perform tracheotomy, and afterwards to divide the thyroid cartilage, by vertical incision. Local anæsthesia having been first produced, Mr. Couper opened the trachea and inserted a tube. After allowing the patient a respite of half an hour, he then very carefully divided the thyroid cartilage in the median line. As, from repeated laryngoscopic examinations, I was so well acquainted with the exact situation of the growth, Mr. Couper requested me to effect evulsion. The larynx, therefore, being held open by means of a powerful retractor on each side, I seized the growth with short, strong forceps, and took the greater part of it away. I then cut the base more cleanly away with some blunt-pointed curved scissors. Mr. Couper finally united the alæ of the thyroid cartilage with three silver sutures.

The patient was relieved by the operation of all dyspnœa, and only complained of some soreness, and slight difficulty in swallowing for a few days. She seemed, to use her own words, "to get well without any trouble." Two of the sutures were removed a week after the operation, but one could not be seen, and sloughed out two months later.

The patient entirely regained her voice, and remained quite well for two years and a half; that is, till the autumn of 1870. An appearance of recurrence began to be visible, with the aid of the laryngoscope, at this time, and the growth slowly increased in the winter of 1870-71. Owing to the irritability of the fauces and nervous condition of the patient, it is impossible to remove the growth from above; and if the neoplasm should attain a large size, the same operation will have to be repeated as was previously done.

CASE LXV.—*Papillary Growth on the Ventricular Band; Evulsion; Cure.*

J. G., æt. 34, a twine-spinner, first attended the Hospital for Diseases of the Throat, May 4th, 1868, on account of complete loss of voice, which had existed for two years.

The laryngoscope revealed a small pink excrescence on the left ventricular band, about the size of a pea. The growth was demonstrated to Drs. Carlill, Sykes, Henry Roberts, and other gentlemen. No attempt at removal was made until June 19th, when the growth was easily seized by common laryngeal forceps.

On microscopic examination, it was seen to consist of papillæ, racemose glands, and some immature connective tissue.

A week later I had the opportunity of showing Mr. John D. Hill, Mr. George Coles, and others, that the larynx was perfectly free: the man already spoke in a natural voice.

CASE LXVI.—*(Papillary) Growth on the Right Vocal Cord; Evulsion; Cure.*

H. E. L., æt. 30, a waiter, was sent to me at the Hospital for Diseases of the Throat, June 4th, 1868, by Dr. Tatham, on account of aphonia, from which he had suffered for twelve months. The voice was generally suppressed, but at times there was a husky sound. There was no dyspnœa.

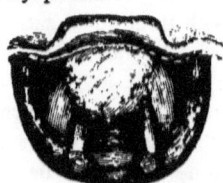

Fig. 61.

On laryngoscopic examination, a white, slightly-furrowed growth, about the size of a winter cherry, was seen occupying the anterior half of the right vocal cord (Fig. 61).

Several pieces were removed at various times, but it was not till July 30th that, in the presence of Mr. Balmanno Squire and Dr. Simpson, of Manchester, I was enabled to remove the whole of the growth with common laryngeal forceps.

The patient came to the Hospital in the following September, and I found the vocal cords entirely free from any irregularity. The voice was in every respect normal and of good power.

CASE LXVII.—(*Papillary*) *Growths on the Right Vocal Cord; Partial Evulsion; Improvement.*

William R., æt. 36, a wool-packer, came under my care at the London Hospital in June, 1868, on account of loss of voice. He had been hoarse for rather more than a year, but had only lost his voice for three months.

On making a laryngoscopic examination, two small growths were seen on the right vocal cord, and on attempted phonation, the vocal cords remained apart posteriorly to the extent of nearly a quarter of an inch. On the first attempt, the whole posterior growth was removed with the tube-forceps; but after repeated operations, the edge of the cord at its anterior half remained rough. The patient recovered a hoarse voice.

In November, 1868, four months after the discontinuance of treatment, the patient remained in the same condition.

CASE LXVIII.—*Papillary Growths on the Right Vocal Cord; Evulsion; Improvement; Recurrence in another part of the Larynx; Evulsion; Recovery of Voice.*

Master W. S., æt. 8, from Maidstone, was recommended to my care by Mr. Sankey, of that town, July 3rd, 1868, on account of loss of voice, which had existed two years. The child's mother stated that his voice first became husky, then gradually hoarse, and finally quite suppressed.

The child had been always humoured a great deal, and was in consequence exceedingly difficult to examine. I managed, however, at the first visit, to obtain a glimpse of the larynx, and to see a warty growth on the right side. Subsequently I ascertained that it was attached to the anterior portion of the right vocal cord (Fig. 62).

The larynx was very small, and the epiglottis very overhanging.

Many weeks elapsed before any instrument could be introduced into the larynx, and the child had to undergo a very laborious

Fig. 62.

training: on each occasion ice had to be sucked for some time before even a laryngeal mirror could be tolerated. He was nearly six months under treatment, and during that time attempts at removal, often unsuccessful, were made almost daily. Ultimately the whole of the growth was removed, but the little boy's voice remained harsh and disagreeable. I saw him from time to time, without noticing any further alteration in the state of the larynx; but in the early part of the year 1870 I observed that there was a fringe of growths on the left vocal cord, and a cauliflower growth on the posterior wall of the larynx (Fig. 63).

Fig. 63.

The latter growth, though itself small, was large in proportion to the size of the child's larynx.

In the course of some months the whole of the excrescences were removed, and the larynx was examined by Mr. James Keene in July, and pronounced to be perfectly free from any growths.

On November 26th, 1870, I received the following report from Mr. Sankey :—

" I am very glad to tell you that he has been steadily improving for the last three months: he is away at school; his voice has returned, although it is not free from huskiness yet, but improves from week to week."

CASE LXIX.—(*Papillary*) *Growths on both Vocal Cords; Tracheotomy and Division of Thyroid Cartilage; Relief of Dyspnœa; Permanent Aphonia.*

Caroline M., æt. 12, was admitted into the Hospital for Diseases of the Throat, July 9th, 1868, but had been treated as an occasional out-patient for several years. The patient had suffered from constant loss of voice, almost since birth, and at intervals from most severe attacks of dyspnœa; and her father stated that several times an operation (tracheotomy?) had been recommended. This case was even more difficult than the last; as, in addition to a very sensitive pharynx, the child had a most violent temper, and frequently would not submit to any treatment whatever. I, however, satisfied myself by repeated laryngoscopic examination, that the larynx was occupied by growths, situated on both vocal cords. It had been hoped that as the patient grew older, she would more calmly submit to treatment; but this idea proving incorrect, and the attacks of dyspnœa becoming more urgent, it was determined, with the consent of the parents, to perform tracheotomy, and afterwards thyrotomy. By this operation, which was performed by my colleague, Mr. George Evans, on July 15th, two warty growths, each about the size of a raspberry, were removed. They occupied the entire length of each cord. The same operation was performed as in Case LXIV., but chloroform was administered. The patient made a good recovery, and she has never since suffered from dyspnœa; but her voice continues to be entirely suppressed. The larynx has been examined several times during the last two years with the laryngoscope; there is no recurrence of the growth, and no congestion of the mucous membrane. The aphonia appears to be due to diminished tension of the vocal cords, which however approximate perfectly.

CASE LXX.—*Papillomatous Growth on the Right Vocal Cord;*
Evulsion; Cure.

Henry A., æt. 48, wool-packer, attended as an out-patient at the Hospital for Diseases of the Throat, July 9th, 1868, on account of loss of voice, which had existed for five years and a half. He had occasionally been able to sound his voice, but never for more than a day or two at a time.

Laryngoscopic examination revealed a pedunculated warty growth, the size of a winter cherry, situated on the middle of the right vocal cord. At the second visit, in the presence of Drs. Jagielski, O'Keefe, Chatterton, and others, I succeeded in removing the whole of the growth, at the first attempt, with common antero-posterior forceps. The tumour was found, on microscopic examination, to be of a papillomatous character.

I saw the patient in January, 1870, and found the larynx free and the voice normal.

CASE LXXI.—*(Papillamatous) Growths on both Vocal Cords ; Treatment by Evulsion ; Cure.*

Mr. S., æt. 27, a telegraph clerk at Malta, was recommended by several medical practitioners at that station to come to England, in order that he might consult me about his loss of voice.

He had been hoarse for three years, but had only suffered from complete loss of voice during the last five months. He had latterly been troubled with shortness of breath, and by a continual hacking cough. His voice was of the greatest importance, as without it he would have been unable to retain his situation.

On laryngoscopic examination, on July 13th, 1868, I found a warty growth occupying the anterior two-thirds of the right vocal cord, and another smaller excrescence in the same position on the left cord (Fig. 64). The one on the right side overlapped that on the left, and it was only on attempted phonation, when the right growth was pressed upwards and outwards, that the outline of the left growth could be seen (Fig. 65). The extent of its attachment below was

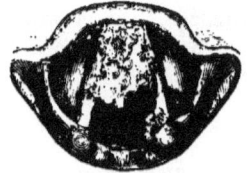
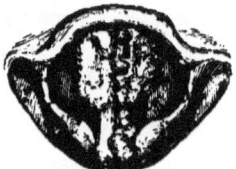

Fig. 64.   Fig. 65.

not ascertained until a later period, when the growth on the right cord had been removed. There was also another small growth, the size of a pea, on the posterior part of the left vocal cord.

I recommended the patient to see me at the Hospital for Diseases of the Throat, where, on July 16th, I removed the whole of the growth on the right side with common laryngeal forceps, in the presence of Dr. Gray, of Oxford, and other gentlemen. After some months, the rest of the growths were removed, and the patient's voice entirely restored.

I saw this patient in September, 1869, and his voice was then perfectly good, and the larynx free from any recurrence of the growth.

CASE LXXII.—(*Papillary*) *Growth on the Right Vocal Cord; Treatment by Evulsion; Great Improvement.*

Thomas R., æt. 47, hawker, attended at the Hospital for Diseases of the Throat as an out-patient, on August 14th, 1868, on account of hoarseness and, at times, almost complete loss of voice. His voice had been bad enough to prevent him from following his occupation for six months, but he had suffered from sore throat for some years. He had formerly had constitutional syphilis. The case was first seen, during my absence from town, by Mr. Lennox Browne, who, on laryngoscopic examination, discovered a cauliflower-like growth, the size of a juniper-berry, on the middle of the right vocal cord. The whole of the mucous membrane of the larynx was thickened and congested. ·

On my return to London, I succeeded in removing the growth with antero-posterior laryngeal forceps, at the second visit of the patient. On account of long-standing chronic hyperæmia, some dysphonia remained, but the man was perfectly satisfied with a loud and rather harsh voice.

CASE LXXIII.—(*Fibro-cellular*) *Growth at Anterior Part of the Larynx; Evulsion; Restoration of Voice; (subsequent) Paralysis of the Abductor of the Left Vocal Cord; Laryngotomy. (Death two years later.)*

Miss Annie S., æt. 18, came from Plymouth to be under my care, in October, 1868, on account of loss of voice, attended with difficulty of breathing. She stated that she had been quite well until two years previously, when "she first noticed that in singing, her voice would go off into a sort of squeak, which would last for a second, and then her voice would become natural again." This peculiarity was not noticed in her ordinary speaking voice till some time later, when it was observed that at times her voice would involuntarily assume a high falsetto tone. Changes of air always had the result of restoring the voice, for the first day after the change. Three months before she consulted me, however, the voice had become uniformly hoarse, and difficulty of breathing also had existed for about the same time. It was first observed at night, when she would wake up in her sleep with a sensation of choking. At first this happened only about once a week, but afterwards it occurred almost every other night. There was no pain in swallowing.

She stated that her father's brother had died of consumption, but that the other members of the family were healthy.

With the laryngoscope, a growth was discovered at the anterior commissure of the vocal cords, and on attempted phonation, it was seen that the neoplasm entirely prevented any movement of the left cord. Some of the growth was removed with tube-forceps, on the 14th of October; but there was great difficulty, on account of considerable enlargement of the right tonsil, and it was found necessary to excise the hypertrophied gland. After this, the remainder of the growth was easily removed on the 1st December. In January, the patient spoke in a clear voice, but the breathing was still difficult, and there appeared some defect in the abductive action of the left vocal cord. It was impossible to tell whether the paralysis had always existed, the position of the growth mechanically preventing its becoming apparent, or whether the loss of power came on subsequently to the removal of the neoplasm.

The stridor increased so much in February and March, that on the 23rd of the latter month it became necessary to perform laryngotomy. This was done by Mr. Evans, and the patient made a good recovery. The abductor of the left cord remained permanently paralyzed, and the patient was unable to dispense with the tube.

This patient came under treatment again eighteen months later, that is, in October, 1870, suffering from severe dysphagia, marasmus, and hectic. She succumbed to these symptoms in the course of a few weeks.

On *post-mortem* examination, there was found to be general disease of the laryngeal cartilages, and a fistulous communication between the pharynx and larynx. There was no sign of the growth.

CASE LXXIV.—*Fibro-Epithelial Growth on the Left Vocal Cord; Evulsion; Relief; subsequent Increase of Growth; Tracheotomy; Complete Evulsion of Growth through Mouth; Cure.*

David L., æt. 41, labourer, came to me at the London Hospital, in October, 1868. He had been hoarse for thirteen years, and had latterly been short of breath. The patient, who was a very nervous, pale, puffy-looking man, admitted that he had been too much addicted to drinking. His fauces were relaxed and sodden. His uvula was elongated, and on laryngoscopic examination, a warty growth was seen attached along the whole length of the left vocal cord, and occupying the whole of the left side of the larynx. It was white and cauliflower-like in appearance. A large portion of the growth was removed on three separate occasions, with common antero-posterior forceps, and the patient's voice improved so much that he discontinued his visits till December, 1869, when he returned with an aggravation of his former symptoms. He stated that he had lately suffered much from a violent cough and spitting of blood. With reference to the latter symptom, he said that on each of the two nights previous to his return to the hospital, he had brought up as much as two or three pints of pure blood. A most careful stethoscopic examination of the chest failed to discover any cause for this hæmorrhage.

Several portions of the growth were removed in March and April, 1870, and the patient discontinued his attendance at the hospital. In August, however, he applied again, with great dyspnœa. He visited the hospital several times, and was seen by my colleagues, Dr. Fenwick and Dr. Woodman. When I returned to town in October, I found his breathing very much embarrassed, and at his urgent desire he was admitted into the Hospital for Diseases of the Throat. On October 10th I removed a small piece of growth with common laryngeal forceps. This was followed by considerable spasm. A few days later, that is on the 14th, I again removed a small piece of the growth. Scarcely had I left the hospital, when such extreme dyspnœa supervened, that it became necessary to perform tracheotomy. This operation was at once successfully done by Mr. Pugin Thornton. Since then, the patient has, of course, been free from dyspnœa, and I have since been able to remove the whole of the growth with my incisive forceps. The canula was removed in April, 1871, and there has been no fresh recurrence of growth. The patient speaks in a clear voice.

The portions of the growth removed were examined by Mr. Stephen Mackenzie, and found to be of fibro-epithelial character. In some parts of the growth, nothing but cells could be found, whilst in others the structure was entirely fibrous, and again, in others, the two elements were blended together (Plate I. fig. 5).

CASE LXXV.—*(Papillary) Growth on the Left Vocal Cord; Evulsion; Cure; Recurrence after a year; Evulsion a second time, and Cure.*

Wm. M., æt. 15, a page, was sent to the Hospital for Diseases of the Throat in November, 1868, on account of hoarseness, which had

23

troubled him for nearly a year. There was no cough nor dyspnœa. On laryngoscopic examination, a growth about the size of a pimento berry was seen to be situated just behind the vocal process of the left vocal cord. The neoplasm was pale, smooth, and sessile. After several ineffectual attempts, I removed the whole of the growth, in the presence of Drs. Greenaway, Gourlay, and other gentlemen. The voice was quite restored. The patient returned in September, 1869, and on examination there was found to be a recurrence of the growth. In colour and size it resembled a millet-seed. On account of its small size, I had great difficulty in removing this second growth, and it was not till the sixth attempt, in the presence of Dr. Stage, of Copenhagen, and Mr. James Keene, that I was successful in seizing it. The boy's voice is now perfect.

CASE LXXVI.—(*Papillary*) *Growth beneath Anterior Insertion of the Vocal Cords ; Treatment by Evulsion ; Cure.*

Mrs. Maria T., æt. 31, was sent to me, at the Hospital for Diseases of the Throat, in December, 1868, by Dr. Bäumler, who kindly supplied me with the following history of her case, from his note-book, sometime after the patient had left my hands.

" *September 23rd*, 1868.—Mrs. T. has always enjoyed good health, except during the last seven months, during which she has almost constantly been very hoarse, and has even, at times, lost her voice altogether. Temporary hoarseness has occurred for the last nine years, but, at intervals, her voice has become quite clear again. Since the hoarseness was permanent, she has gradually become weaker. At the time her voice first became affected her hair came out very much, but there was no sore throat nor any eruption in the skin. She had been married two years, and has one child. Eleven years ago she had an inflammation of the eyes, which required scarification. (Cicatrices and slight synechiæ are still visible in the conjunctiva of the left eye.)

" Present state, rather anæmic, although for the last two months she had taken mist. ferri perchlor. Voice without sound, but at times a somewhat clear note.

" Velum palati and pharynx normal. Larynx also quite normal, with the exception of the anterior extremity of the right vocal cord and the adjoining anterior commissure, where a warty excrescence, the size of a lentil, is visible. It seems partly to grow out of the vocal cord, and partly from the commissure (Fig. 66). On phonation it partly lies between the anterior part of the vocal cords, and

Fig. 66.

prevents their proper approximation. That part of the right vocal cord which adjoins the excrescence is a little red.

"*January 7th*, 1869.—Has been in the Hospital for Diseases of the Throat, and operated upon by Dr. Morell-Mackenzie. Voice still somewhat husky. The right vocal cord is somewhat injected; the prominent part of the growth has been removed, and only its broad base is now visible."

I have only to add to this careful summary, that when the patient was admitted into the Hospital I found another larger growth beneath the anterior insertion of the right vocal cord. The case was a difficult one for observation, and still more for treatment; but after being trained for a few days, the patient was able to submit to operative procedures, and the growth was so far removed that a slight unevenness only remained at the anterior surface of the cord. The lower growth was entirely removed.

On the 24th of March, 1869, a laryngoscopic examination was made by Mr. Lennox Browne in my absence, who reported that "the larynx was quite normal, and that the voice was clear and resonant."

CASE LXXVII.—*Papillary Growth on the Posterior Wall of the Larynx; Evulsion; Cure.*

Henry W., æt. 27, a vocalist, applied as an out-patient at the London Hospital, in January, 1869, on account of hoarseness in ordinary speaking, and total loss of power in singing. He also complained of a cough, which was unaccompanied with expectoration. On laryngoscopic examination, a growth the size of a blackberry was seen springing from the posterior wall of the larynx. It was slightly irregular in outline, and of pink colour. The patient discontinued attending till May, when I had an opportunity of showing him to my class. In June the whole of the growth was removed at three sittings, with tube-forceps. The particles removed were found by Dr. Fenwick to consist of cylindrical papillæ. In the following September the patient applied again at the Hospital on account of slight rheumatism. On laryngoscopic examination, the larynx was found to be perfectly healthy.

CASE LXXVIII.—*Fibrous Growth on the Left Vocal Cord; Treatment by Evulsion ; Cure.*

Mr. B., æt. 42, a publican, consulted me in February, 1869, on account of shortness of breath and loss of voice. The symptoms were found, on laryngoscopic examination, to be caused by a large whitish, cauliflower, growth situated on the left vocal cord. Finding that the patient, who was a German, had formerly been under the care of Dr. Hermann Weber for another complaint, I proposed a consultation with that gentleman, and it was finally agreed between us, that I should attempt removal with the aid of the laryngoscope. Owing to the extremely irritable condition of the pharynx, I scarcely expected to be successful, and the probable necessity of division of the thyroid cartilage was discussed at the consultation.

During the year 1869 I managed to remove several large pieces, and the patient became able to speak in a hoarse voice, and was entirely relieved of his dyspnœa. He was still seen by me occasionally in April, 1870, and there remained a little irregularity of the vocal cord. The ordinary laryngeal forceps, as well as those opening in the antero-posterior direction, and the tube-forceps, were used in this case.

Since writing the foregoing, I have succeeded in removing the last fragment, and in October, 1870, the larynx was perfectly healthy, and the voice quite natural. The following microscopic report of pieces removed, April 23rd and 30th, was furnished me by my brother, Mr. Stephen Mackenzie :—

"The growth submitted for examination was about the size of a dried pea, somewhat square-shaped, of pearly whiteness, and weighing one decigramme.

"One side appeared to be that by which it was attached to the vocal cord—here it was marked by blood. This surface was smooth and uniform. Springing from its opposite side were a number of filaments, or fringe-like processes, packed very closely together.

They were of about uniform thickness and length, quite white, and very tough, so that they were only detached from the surface from which they sprang with great difficulty ; a nail-brush would convey a very good idea of this growth, the ivory part representing the base from which the processes sprang, and the bristles the filamentous processes themselves (Fig. 67).

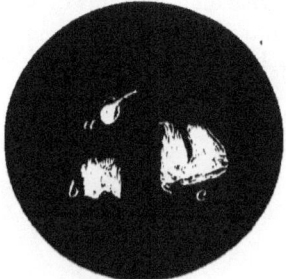

Fig. 67.

" *a* and *b* refer to the pieces of growth removed on April 23rd ; *c* to portion removed on April 30th."

Examined under the microscope, there was found to be nothing but white fibrous tissue, very dense in composition (Plate I. fig. 3). No trace could be found of epithelium, elastic tissue, cells, or vessels.

CASE LXXIX.—*Adenoma beneath the Anterior Insertion of the Vocal Cords ; Treatment by Evulsion ; Cure.*

Alfred S., æt. 33, a stonemason, first applied at the London Hospital early in March, 1869, giving the following history :—

Two years previously he had become hoarse, but thinking that this was due to a cold, he paid no attention to it ; the voice, however, gradually became thicker and more husky. He suffered no pain, but was troubled with a constant dry cough. Nine months previously he had applied at the Victoria Park Hospital, where he had attended as an out-patient for three months. He had never spat blood, but had suffered from occasional attacks of dyspnœa, with stridor; he had not had syphilis, and his family history was good.

On laryngoscopic examination, a small growth was observed beneath the anterior commissure of the vocal cords (Plate III. fig. 6). This growth proved a very difficult one to extirpate, as the vocal cords became spasmodically approximated when any instrument was introduced within the larynx. A great variety of instruments were accordingly used. The first pieces were removed with the tube-forceps; at a later period I used the common antero-posterior laryngeal forceps, and finally a portion was taken away with Stoerk's wire guillotine.

On microscopic examination, the growth was seen to be almost entirely made up of an hypertrophied racemose gland, two sections of which are shown in Fig. 68. The time over which treatment

Fig. 68.

extended was twenty weeks, and whilst under treatment he was seen by the Rev. Dr. Bell, M.D., and many medical practitioners. I have examined the larynx lately (April, 1870), and it is entirely free from any recurrence. The voice is perfectly clear, and there is no dyspnœa.

CASE LXXX.—*Symmetrical (Papillomatous) Growths on the Vocal Cords; Treatment by Evulsion; Cure.*

Mr. W., æt. 60, a retired merchant, living in Manchester, consulted me April 21st, 1869.

The previous history of the patient was that he had enjoyed tolerably good health until a year previously, when he had seen me, on account of hoarseness and a troublesome cough. I had then discovered nothing but congestion of the larynx and trachea, with a tendency to bronchitis. He had derived great benefit from local treatment, and had left me quite well. When, two months later, he wrote, telling me that he was suffering from his former symptoms, I advised him to place himself under the care of Dr. Simpson, of Manchester. That gentleman soon informed me that he noticed two growths on the vocal cords, and by my advice, for some weeks, he applied caustic solutions to the excrescences. This treatment proving unavailing, Dr. Simpson recommended the patient to come to town to see me.

On laryngoscopic examination, I found a small wart, about the size of a coriander-seed, symmetrically situated on each vocal cord, immediately behind the vocal process (Fig. 69). From the small size of the growths, and the fact that introduction of

Fig. 69.

any instrument generally caused violent cough, some difficulty was experienced in seizing them. At the end of a few weeks, however. I effected evulsion with tube-forceps, and all the symptoms speedily disappeared.

CASE LXXXI.—(*Papillary*) *Growths on the Posterior Wall of the Larynx below the Glottis, with Paralysis of the Abductors of both Vocal Cords ; Laryngotomy and subsequent Removal of Growth through Tracheal Opening ; Cure.*

Mary Ann D., æt. 51, a charwoman, applied at the Hospital for Diseases of the Throat, May 27th, 1869, on account of extreme difficulty of breathing, which had existed in a severe form for two years.

She stated that thirty-six years previously she had suffered from a severe attack of measles, and that at the time her throat had been much affected, and had been lanced two or three times. Ever since her voice had frequently been suppressed during the winter months. Occasionally she was only hoarse ; at other times she could hardly whisper. These attacks sometimes only lasted for a week, but once she lost her voice for six weeks. Twenty-nine years ago she had small-pox, and since that time had never been strong; she had, however, married, and had brought up a family of six children.

On admission, she was suffering from embarrassed and stridulous respiration ; she had a croupy cough, with scanty expectoration, and her voice was almost entirely suppressed ; she did not complain of pain, but of an occasional feeling of choking. She was very thin and haggard, and looked at least ten years older than her age. On making a laryngoscopic examination, it was seen, that, on inspiration, the vocal cords were very little abducted from the median line, but remained nearly approximated (Fig. 70). The larynx appeared otherwise quite healthy, and the vocal cords were perfectly white. It was judged to be a case of paralysis of the abductors, and the symptoms being somewhat urgent, laryngotomy was at once performed (May 30th). The patient experienced immediate relief when the canula was inserted.

On June 7th, on making a laryngoscopic examination, it was observed for the first time that a small portion of a growth projected upwards, between the vocal cords (Fig. 71). This con-

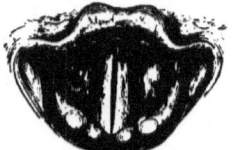

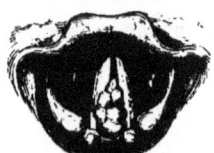

Fig. 70.   Fig. 71.

dition having been verified on two or three occasions, an attempt was made to seize the growth with short curved forceps, introduced through the wound (the canula having been previously removed). The first trial was not successful; but on the 8th of June a growth ¾ of an inch in its long diameter, by 1½ inch in circumference, was removed. The patient was discharged, wearing the tube, on the 19th of July. She applied again in November, and on December 3rd she was re-admitted: the abductive action of the vocal cords being now perfect, and there being no trace of recurrence of the growth, the canula was removed. She suffered no inconvenience, made a good recovery, and is now (June, 1870) quite well, the respiration and vocalization being in every respect natural.

In this case the growth appears in some way to have acted mechanically, and thus to have prevented the vocal cords being abducted by the crico-arytenoidei postici. The small space between the cords accounts for the circumstance that the growth (which was situated beneath them) was not seen till after the conditions had been altered by the opening in the windpipe.

This patient during treatment was seen by Dr. Boddaert, of Ghent, Mr. Wordsworth, and many medical practitioners.

CASE LXXXII. —*Papillomatous Growths covering almost the entire Mucous Membrane of the Larynx; Treatment by Evulsion; Cure; Recurrence; still under Treatment.*

Rebecca L., æt. 21, was admitted into the London Hospital, on account of great difficulty of breathing, May 14th, 1869. Her mother stated that she had never been able to sound her voice, and that when she was a year old, it was noticed that she cried very hoarsely. The shortness of breath began to come on a year ago. She coughed up a great deal of phlegm, and made such a loud croupy noise in her sleep that no one could rest in the room with her. She first attended at the Victoria Park Hospital, and was afterwards in the Metropolitan Free Hospital for six weeks. Later still she was an out-patient at the latter institution, and had attended there till within six weeks of her admission into the London Hospital. On admission, the patient was found to be suffering from considerable dyspnœa and stridulous breathing, and it was thought that tracheotomy would be necessary. She passed a very bad night, being obliged to sit up all the time, but obtained some relief towards the morning from benzoin inhalations.

The next day I saw her for the first time, and on laryngoscopic examination I found growths on both vocal cords, entirely occluding both cords and ventricular bands. The larynx was exceedingly small, not larger than that of a child eight years old, and I had the greatest difficulty in introducing any instrument into the larynx. I however succeeded with tube-forceps, on the first* occasion, in removing several pieces, which, in the aggregate, were as large as a pea. After some weeks all the growths had been removed, with the exception of a very small piece on the right vocal cord. This last remnant proved a great difficulty; the pharynx was so sensitive that the patient was obliged to suck ice for nearly an hour before she could be operated on; the laryngeal aperture was, as observed, extremely small, and the base of the growth proved to be very hard. However, at the latter end of April the voice commenced to be distinctly phonetic—the dyspnœa had disappeared long since—and in May I had the satisfaction of seeing that the larynx was entirely free of growths, and of hearing a clear and even agreeable voice. In the spring of 1870 a slight recurrence of growth took place on the right ventricular band, and the patient is still under treatment. Portions of the growth removed were examined under the microscope by my brother, Mr. Stephen Mackenzie, who reported that they consisted of large papillæ covered with several layers of squamous epithelium (Plate I. fig. 1).—(Case briefly reported in *Lancet*, vol. i., 1869, page 229.)

CASE LXXXIII.—*Papillomatous Growths on the Right Vocal Cord and Under Surface of the Epiglottis; Partial Evulsion; Improvement.*

Eliza B., æt. 51, came under my care at the London Hospital, May, 1869, on account of hoarseness and dysphagia, which had existed for four years.

With the aid of the laryngoscope, it was seen that there were two growths in the larynx : one, the size of a large pea, was attached by a broad base on the right vocal cord ; the other, not larger than a grain of wheat, grew from the under surface of the epiglottis. The patient had never had syphilis, and there was no evidence of phthisis.

The growth on the epiglottis was removed at the second visit with common forceps, and the greater part of the growth on the right vocal cord a few days later. These portions, examined by Dr. Fenwick, were pronounced to be simple warty growths. The result of treatment is unknown in this case, as the patient discontinued attendance after the second visit. There can be little doubt but that great improvement must have taken place.

CASE LXXXIV.—*Papillary Growth on the Left Vocal Cord ; Treatment by Evulsion ; Cure.*

Emma P., æt. 21, servant, was sent up to the London Hospital from Maidstone, in June, 1869, on account of great difficulty of breathing. She had had three attacks of strangulation, each of which had come on during the night. The first attack occurred in December, 1868. She could only lie down on her right side, and was obliged to sleep with her hand under her neck. If, whilst asleep, her hand slipped away, she always woke with difficulty of breathing.

Occasional loss of voice had been noticed as early as October, 1867, and from that time till the following April she could speak hoarsely by making a great effort. Since April, 1868, she had been unable to sound her voice at all.

On laryngoscopic examination, a smooth, pale, growth was seen to be growing from the anterior two-thirds of the left vocal cord. It appeared to be as large as a common hazel nut, and at the first visit a piece was removed, which was believed to be the whole growth. In the following week, however, it was seen that a piece as large as a gall-nut still remained attached to the under surface of the left cord, and in addition, there was a growth the size of a small pea attached to the centre of the right cord. A large piece was again removed, and subsequently, as many as a dozen pieces have been taken away. Tube-forceps, common forceps, and Stoerk's wire-guillotine were all used in this case. It is worthy of note, that this case was one of those rare instances in which pain attended the operation. The patient always complained of severe pain in the ears after each attempt at removal.

Although the symptoms greatly improved, the progress of the case was for some time very disappointing. No sooner was one neoplasm removed, than another, situated lower down, was revealed. In August, 1870, the right cord was quite clear, but a small portion of growth remained low in the larynx, beneath the left cord. The patient had entirely lost all symptoms of dyspnœa, and her voice was quite clear.

On December 13th, 1870, I saw this patient after an interval of nearly four months. The remnant on the left cord appeared to be hanging by a mere thread, and was easily removed. I had the opportunity of showing her a few days later to Dr. Elsberg, of New York, who examined the larynx, and agreed with me that it was then quite clear ; the patient spoke in a perfectly natural voice.

CASE LXXXV.—*Cystic Tumour on the Epiglottis ; Puncture and Evacuation of Contents ; Cauterization ; Cure.*

M. W. W., æt. 22, farrier, applied at the Hospital for Diseases of the Throat, June 28th, 1869, on account of difficulty of swallowing, and slight hoarseness. He stated that he felt a constant desire to clear his throat, which seemed always full of phlegm. Swallowing was not painful, but difficult ; and food, especially fluids, constantly went the wrong way. His voice was slightly husky. He stated that he had been constantly exposed to inclement weather, and that his symptoms had existed about six years.

On examination with the laryngoscope, a large globular swelling, the size of a cherry, was seen on the upper surface of the epiglottis (Fig. 72, and Plate III. fig. 2). This swelling projected through the under surface of the epiglottis, so as to almost entirely hide the right half of the larynx (Fig. 73, and Plate III. fig. 1). A few days later,

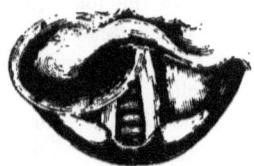

Fig. 72.   Fig. 73.

I made a free opening in the tumour with my unguarded laryngeal lancet, in the presence of Dr. Gore Ring and others. There was a free discharge of steatomatous-like matter, with a small quantity of glairy fluid and blood. On July 26th there was not the slightest sign of the tumour, nor even of the puncture.

CASE LXXXVI.—(*Papillomatous*) *Growth at the Anterior Commissure of the Vocal Cords; nearly Complete Evulsion; Great Improvement.*

Mr. J. Y., æt. 43, a farmer, was sent to me by Dr. Lewis, of Basingstoke, in September, 1868, on account of an altered voice, and an extremely disagreeable sensation in the windpipe.

A laryngoscopic examination showed considerable congestion of the lining membrane of the larynx, and diminished mobility of the vocal cords. Local applications, of an astringent character, were repeatedly made to the throat for five weeks, and the patient left, having obtained, as he stated, considerable relief.

Rather less than nine months afterwards, however, he applied to me again, on account of loss of voice and frequent attacks of suffocation; and on making a laryngoscopic examination, I discovered a large, irregular, lobulated growth, about the size of a raspberry, beneath the anterior commissure of the vocal cords, blocking up half the space of the glottis. Notwithstanding that the patient was an exceedingly nervous man, and did not tolerate the laryngoscopic examination at all readily, in the course of a month I removed several pieces of growth; in the aggregate these would have made up a growth about the size of a large cherry. They were taken away partly with tube-forceps, and partly with Stoerk's écraseur. At the end of July, when I left town for my vacation, there was still a small portion of growth remaining, but the patient was able to breathe without difficulty, and could sound his voice. In January, 1870, I heard from him to the following effect : " My voice is still very weak, but I can talk a little in the *natural voice.*" (The italics are his own.)

CASE LXXXVII.—*Epithelioma on the Right Vocal Cord ; Tracheotomy and Thyrotomy ; Constant Development of Growth ; Death.*

Mr. John S., æt. 47, silversmith, consulted me September 18th, 1869, on account of hoarseness, attacks of suffocation, and slight difficulty in swallowing. The patient was a stout, thick-set man, but his countenance was pale and anxious, and his pulse feeble. He stated, that for the last five years his breathing had been short, but he attributed this symptom to his increasing stoutness. Four months previously, however, he had been alarmed by a violent attack of suffocation during the night. These attacks came on more frequently, and a month later his voice became hoarse. He also suffered from most violent paroxysms of cough. His digestion was not good, but his appetite was almost voracious ; in fact, he really ate as much as two or three ordinary men. After his meals, his difficulty of breathing was always much increased. He had received some general treatment, but without relief. On examining the throat, the pharynx was seen to be greatly relaxed, and the uvula elongated. The fauces were so irritable, that a mere inspection even of these parts caused an inclination to retch. After some difficulty, however, I

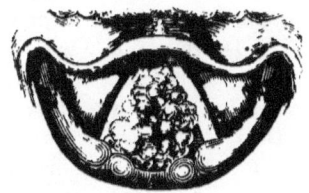

Fig. 74.

succeeded in making a successful examination of the larynx, and discovered a red cauliflower-growth, the size of a cherry, projecting from the right vocal cord (Fig. 74) : the mucous membrane generally was congested and relaxed. It was only after seeing the patient several times, that he permitted me to make any attempt at removal of the growth, and even then he was so nervous and restless, that my efforts were unsuccessful.

On the third occasion, however (September 27th), I succeeded in removing a portion about the size of a large pea. The piece was removed with very slight force—not more than is commonly necessary in eradicating soft neoplasms growing on mucous membrane, but the patient complained of extreme pain, and on this account obstinately refused any further laryngoscopic treatment.

Two days later (29th) Mr. S. took cold, and his breathing became much worse. As the symptoms were urgent, and the patient was willing to have an operation performed under chloroform, tracheotomy and thyrotomy were proposed and acceeded to.

Accordingly, on October 3rd, Mr. Wordsworth opened the trachea. The operation was attended with considerable difficulty, on account of the shortness of the neck and the extraordinary amount of fat in front of the windpipe. In addition to this, the patient bore chloroform very badly, and showed marked signs of cardiac syncope. On this account, it was not thought safe on that occasion to perform thyrotomy. The patient did fairly well after the operation, and a fortnight later a vertical incision was made through the thyroid cartilage, and several large pieces of growth were removed from the larynx ; the parts were afterwards brought together with silver sutures. The wound healed, but the patient was advised to continue to wear the tube for a time. As the immediate result of the operation, it may be stated that the dyspnœa entirely ceased, that the swallowing was effected with ease, and that the voice, though hoarse, became for a short time distinctly phonetic. The patient, however, continued to suffer from attacks of coughing, which he did not at all attempt to restrain.

At the beginning of December, it was noticed that the weak granulations, commonly seen beneath the shield of a canula, when a tube has been worn for a few weeks, were larger and more numerous than is usual. This condition was thought to be sufficiently accounted for by the fact that the patient was so penurious, that he not only refused proper—that is, sufficiently frequent—medical attendance, and nursing, but he even grudged clean linen and changes of clothing. Thus it happened that there was always a certain amount of moisture around the canula. In spite of the application of solid nitrate of silver to the exuberant papillæ, and the frequent use of powdered oxide of zinc, and of various other remedies, the vegetations extended in all directions. He began also to show signs of depression, though his appetite was still good, and he slept fairly well.

Laryngoscopic examinations had been made from time to time, and in January, 1870, it was first observed, that there was some recurrence of the growth. In the beginning of March the neoplasm was seen to rise above the level of the ary-epiglottic folds, and to block up the entire larynx (Fig. 75). The patient now suffered from

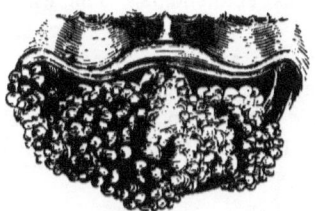

Fig. 75.

frequent suffocative attacks of coughing, apparently on account of the tube becoming obstructed. Repeated examinations of the canula during these attacks showed that it was not blocked up by mucus, and it was judged that the growth had extended downwards, and obstructed the lower end of the canula. The patient died during one of these attacks in the night of May 10th, 1870.

On making a *post-mortem* examination, it was found that the whole of the interior of the larynx was blocked up by an enormous cauliflower excrescence, which extended from the level of the ary-epiglottic folds, downwards, for more than four inches, thus reaching quite an inch below the orifice in the trachea, made in tracheotomy. It likewise penetrated to the base of the right ary-epiglottic fold, and had extended into, and enlarged, the right hyoid fossa (Plate IV. figs. 2 and 3). The growth extended along the track of the canula to the outside of the neck, and formed a thick fringe, an inch and a half in width, around the orifice of the tracheal opening; so that the supposed granulations under the canula shield proved to be neoplasms of exactly similar formation to the growth in the larynx (Plate IV. fig. 1). The luxuriant growth of the new formation in this case pointed to its being otherwise than of benign character, and its microscopic examination illustrates the extreme difficulty of arriving at accurate conclusions concerning the histology of these tumours, even when the entire growth is brought under observation. The specimen was examined by several eminent microscopists, and was at first reported to be a simple papilloma. On another examination, fibrous tissue was found to be largely developed, and it was pronounced fibro-cellular. Still later, my brother, Mr. Stephen Mackenzie, discovered some nested cells (laminated capsules of Paget). One of these is depicted in Plate I. fig. 8, and from the extreme importance of this element, the case must undoubtedly be placed in the category of carcinomatous growths, and be considered as epithelioma. The whole surface of the growth was covered by papillomata. The fact that during life it was believed to be, and was treated as, benign, had led me to include it amongst these cases; and having done so before repeated microscopic examinations had succeeded in discovering any cancer elements, I did not think it fair to exclude it afterwards.

CASE LXXXVIII.—*Adenoma on the Under Surface of the Epiglottis; Tracheotomy; Removal of Growth by Écraseur; Recovery.*

A retired Indian officer, æt. 51, applied to me on October 12th, 1869, on account of an exceedingly troublesome cough, slight hoarseness, and occasional dysphagia. The symptoms had commenced about a year before. Twenty years previously the patient had suffered from syphilis. On making a laryngoscopic examination, there was found to be superficial ulceration, with slight thickening of the left side of the epiglottis (Fig. 76). The ulceration yielded rapidly

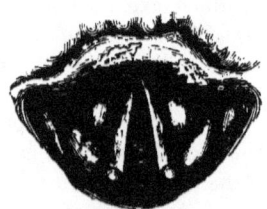

Fig. 76.

to treatment, consisting of iodide of potassium and the local application of mineral astringents. On discontinuance of the remedies, however, there was a great disposition to recurrence, and the patient exhibited a decided catarrhal tendency. Under these circumstances, on November 6th, I recommended him to pass the winter at

Cannes. I received favourable reports of him for some weeks, but early in January, 1870, I heard from Dr. Frank that a laryngoscopic examination, made by himself and Dr. Wagner, of Königsberg, had discovered "a large sarcomatous neoplasm" on the under surface of the epiglottis. These gentlemen recommended his immediate return to England, in order that the growth might be removed. On January 9th, 1870, I made a laryngoscopic examination, and found an irregularly mammillated growth, of a pale colour, about the size of a cherry (Fig. 77, and Plate III, fig. 4). On

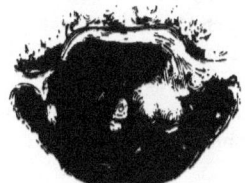

Fig. 77.

several occasions lately, the patient had suffered from severe dyspnœa. I therefore thought it advisable to have tracheotomy performed before attempting to remove the growth, and the windpipe was opened on the 14th of January by Mr. James Adams.

The patient took cold soon after the operation, and suffered from rather severe bronchial catarrh. It was not therefore till March 6th that I proceeded to extirpate the tumour. Chloroform was administered by Mr. Clover, and the growth was removed by means of my wheel écraseur. The tumour weighed fifty grains, and was of the dimensions given in the annexed drawing (1 inch by three-quarters of an inch (Fig. 78, and Plate III. fig. 5). This specimen was ex-

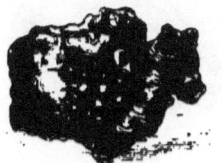

Fig. 78.

hibited at the Pathological Society, and referred for investigation to the Morbid Growth Committee. The Sub-Committee appointed to examine the specimen, considered it a case of "adenoid carcinoma" (*Medical Times and Gazette*, July 16th, 1870), but their

25

report was not confirmed by the full committee, and does not appear in the *Transactions*, vol. xxi. p. 51.

On the next day it was seen that the epiglottis was quite clear, except a small sloughy surface at the outer angle. The swallowing improved immediately, and the voice was also much better. Ten days later I noticed a slight irregular ulceration on the ventricular bands, but this rapidly yielded to treatment. In May, 1870, the breathing being good and the voice natural, the tube was removed, and the patient shortly afterwards went down to the seaside. Unfortunately, however, at the end of a month, he took a severe cold; acute laryngitis supervened, and it became necessary to perform tracheotomy a second time. The patient still wears the tube.

The pathological interest of this case depends on the extreme rapidity of the production of the neoplasm, and on its possible dependence on the syphilitic dyscrasia.

CASE LXXXIX.—*Vascular Tumour in the Right Hyoid Fossa;
Treatment by Evulsion; Cure.*

Captain V., æt. 35, of the Royal Marines, was sent to me in December, 1869, by Dr. Smyly (of Dublin), a well-known laryngoscopist, who had previously examined him with the laryngeal mirror. The patient complained of an uneasy tickling sensation in the throat, which had come on in the previous summer, when he had hay fever. The voice was normal. On laryngoscopic examination, a growth, the size, colour, and configuration of a ripe blackberry, was seen in the right hyoid fossa (Fig. 79, and Plate II. fig. 12). Several

Fig. 79.

attempts were made at removal, but, owing to the extreme hardness of the growth, it was not entirely removed till January 21st, 1870. Incomplete evulsion was effected with tube-forceps, and the base of the growth was excised with cutting forceps.

The operation was attended with more hæmorrhage than is usual, and the patient informed me that for some hours after leaving my house slight bleeding took place. The growth, on removal, had lost its black colour, and was of a rather bright red hue. On section, its appearance to the naked eye was compact, and its colour almost white.

On microscopic examination, it appeared to consist of fibrous tissue of an exceedingly close and matted character. No blood-vessels could be discovered, nor did it contain any blood; but numerous elongated nuclei were seen. The appearance indeed closely corresponded with the description of a case of "venous vascular tumour" removed from the thigh of a patient by Sir William Lawrence. Mr. Paget's remarks (*Op. cit.*, p. 583), in the case referred to are so entirely applicable to my specimen that I feel I cannot do better than quote them :—" The obscurity of the microscopic appearances was due to the tenacity with which the blood-vessels were imbedded in the elastic fibrous or nucleated tissue ; it seemed impossible to extricate complete vessels ; and one, obtained by dissection, only fragments of their walls, confused with the intermediate tissues."

CASE XC.—*Papillary Growths above and below the Anterior Commissure of the Vocal Cords ; Treatment by Evulsion ; Cure.*

Thomas W., æt. 44, labourer, presented himself as an out-patient at the Hospital for Diseases of the Throat, January 20th, 1870, on account of hoarseness, which had existed nearly twelve years. He felt no pain, but always had a sense of tightness about the throat. He had had syphilis fifteen years previously.

On laryngoscopic examination, both vocal cords were observed to be in a state of chronic inflammation, and their edges were very

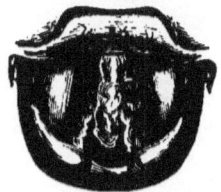

Fig. 80.

irregular. From the anterior commissure a long narrow growth hung down between the cords. It was almost white, and somewhat resembled a nasal polypus in appearance, though less transparent (Fig. 80).

Repeated operations, resulting in removal of one or more pieces, were made on various occasions. Once in the presence of Dr. Thorowgood, a piece five-eighths of an inch long, and one-eighth in breadth, was removed ; but at each succeeding visit, there seemed to be as much growth left as before. This was on account of successive portions of growth becoming visible below the glottis, as other portions were removed by evulsion. Altogether as many as 22 pieces were removed, and ultimately the whole growth was eradicated ; after astringents had been applied for some weeks, the chronic congestion was greatly diminished, and the patient recovered a good voice.

Microscopical examination showed that the growth consisted of the ordinary elements of a benign papilloma.

Case XCI.—*(Papillomatous) Growths on Both Vocal Cords;*
Treatment by Evulsion ; Cure.

Henry C., æt. 47, a wine-bottler, attended as an out-patient at the Hospital for Diseases of the Throat, on account of loss of voice, of six months' standing.

On examination with the laryngoscope, a large growth was seen, occupying the whole length of the right vocal cord, and another smaller one on the left cord. Both cords were very irregular (Fig. 81).

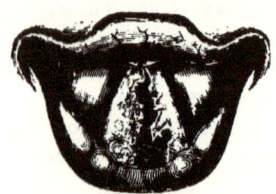

Fig. 81.

The patient was exceedingly nervous, but his employers were good enough to allow him wine *ad libitum* before each visit to the hospital. With the aid of this stimulant, and ice for half an hour before the operation, he was able to undergo treatment. I succeeded very early in the history of the case in removing the larger portion of the growth from the right vocal cord, Surgeon-Major Wyatt, Dr. Stage, of Copenhagen, and Mr. Edwin Peacock, of Birmingham, being present. Common antero-posterior forceps were used.

The growths on the left vocal cord were more difficult to remove, and many attempts were unsuccessful. Finally, however, on May 5th, 1870, the larynx was examined in the presence of Dr. Farquhar, and seen to be clear ; but the voice, though phonetic, still remained slightly gruff. This was no doubt due to the long-standing hyperæmia of the mucous membrane of the larynx. In this case there were in all ten operations, and six pieces were removed, amounting *en masse* to a growth the size of a cherry.

CASE XCII.—*Benign Epithelial Growth on and below the Left Vocal Cord ; Treatment by Evulsion ; Cure.*

Anne M., æt. 48, haberdasher, attended the Hospital for Diseases of the Throat on February 3rd, 1870, by recommendation of Mr. Kennedy, of Stratford. She stated that she had suffered "from loss of voice on and off for twenty-four years." The origin of the aphonia was supposed to be exposure in a snow-storm ; for on the following day she lost her voice, and since that time it has never properly returned. Latterly it had been completely suppressed.

On laryngoscopic examination, it was seen that the left vocal cord and the mucous membrane beneath it, for half an inch, were covered by an uneven growth, which, though occupying a large area, did not appear to project more than an eighth of an inch beyond the level of the mucous membrane.

It was some time before I was enabled to remove any portion ; but on April 25th, in the presence of Dr. Haden, as well as those other gentlemen who were attending the practice of the hospital, I succeeded with forceps in scraping away and detaching an irregular growth, about the size of a split almond. Three days later I found that the growth had been completely removed. There still, however, remained considerable hyperæmia of the left vocal cord.

On microscopical examination by Mr. Stephen Mackenzie, "the growth was found to consist of simple epithelial structure. It was made up of cells irregular in size and shape, some being round, others oval, caudate, and fusiform. The nuclei were very distinct, and most of them large ; some cells contained more than one nucleus, and a few of them were undergoing dichotomous division. All the cells were very granular."

The patient has now (September, 1870) a good voice, though it is not perfectly clear.

CASE XCIII.—(*Papillomatous*) *Growth on the Left Vocal Cord ; Treatment by Evulsion ; Cure.*

Robert W., æt. 30, engine-driver, applied as an out-patient at the Hospital for Diseases of the Throat, March 3rd, 1870, on account of hoarseness, which had existed for eighteen months. He attributed his symptoms to exposure to draughts, and sudden changes of temperature.

On examination with the laryngoscope, a smooth but slightly mammillated growth was seen attached to the left vocal cord. It was easily removed at the second visit, on March 7th, in the presence of Dr. H. Hubbard and others. On April 21st, there having been no return of the growth, and the voice being perfect, the patient was discharged cured. In July he was examined by Mr. Keene, who agreed with me that the larynx was quite healthy.

CASE XCIV.—(*Papillary*) *Growth on the left Vocal Cord; Treatment by Evulsion ; Cure.*

Mary Anne R., æt. 35, a gilder's wife, attended at the Hospital for Diseases of the Throat, March 7th, 1870, on account of loss of voice.

She attributed the aphonia to exposure to cold five years previously. Since that time she had been frequently hoarse for two or three months at a time ; and during the last year her voice had been entirely suppressed. For the last ten weeks she had also suffered from shortness of breath.

On examination with the laryngoscope, a smooth, pea-like growth was seen, attached by a broad base to the left vocal cord. On March 24th it was entirely removed with common antero-posterior forceps, in the presence of Mr. Keene and Dr. Stage, of Copenhagen.

A fortnight later, the voice was natural, and the larynx, examined by each of the above-named gentlemen, was seen to be quite free from any sign of growth.

CASE XCV.—*Sarcomatous Growth on the left Ventricular Band ;
Treatment by Evulsion ; Cure.*

Susan W., æt. 43, a coachman's wife, attended as an out-
patient at the Hospital for Diseases of the Throat, March 7th, 1870,
on account of complete loss of voice.

She stated that she had been subject to sore throat whenever she
took the slightest cold for the last twenty-three years, and that her
voice had been affected, with but very slight intermissions, during
the whole of that period. For the last year, she had been permanently hoarse, and for the last four months her voice had been
entirely suppressed. She complained of soreness and a pricking
pain, " as if a bone were in her throat." Swallowing was painful,
and her breathing was short on the least exertion. Lately, she had
awoke in the night with attacks of choking.

On laryngoscopic examination, a large irregular growth was seen,
on inspiration, to be attached to the right ventricular band, entirely
occupying the right half of the larynx, with the exception of a very
small portion of the right vocal cord, at its anterior insertion (Plate III.
fig. 7). On attempted phonation, the growth projected across the
left vocal cord, and occupied two-thirds of the area of the larynx
(Plate III. fig. 8). The colour of the growth was red, like that of the
surrounding mucous membrane.

On the 17th of March the first attempt at removal was made,
and on this occasion common laryngeal forceps were introduced
into the larynx three times. On the first introduction of the instrument, one large piece was removed, and subsequently two smaller
fragments (Plate III. fig. 9), and the voice became at once per-
ceptibly better : it was, however, impossible on this occasion to see
the condition of the larynx, on account of the slight hæmorrhage
which took place.

On the 24th the voice was perfectly phonetic, and the larynx
quite free of growth, the only remaining symptoms being a
slight irregularity of the right ventricular band, at the posterior
part of the larynx, and some congestion of the right vocal
cord. This subsided, and a week or two later, the patient was
discharged, cured.

This case, the facts of which are known to Dr. Carre, of Black-
heath, was frequently examined, both before and after treatment, by

Dr. Stage, of Copenhagen, Mr. Keene, Mr. Peacock, and other gentlemen.

Microscopical examination of this growth by Mr. Stephen Mackenzie showed that " it consisted of extremely long fusiform cells. Some of the cells were oat-shaped, and some more spherical ; but all had a tendency to elongation (Fig. 82). The cells, for the most part,

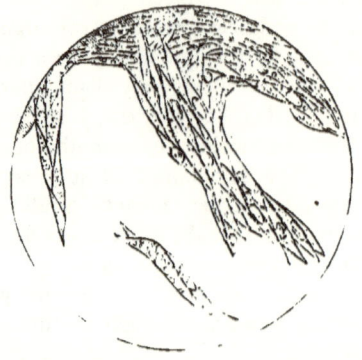

Fig. 82.

possessed pretty distinct, round or oval nuclei, and the cell-contents were granular, and approached a fatty condition ; but it should be mentioned that the specimen had been immersed in spirit for a few days before it was examined. In some parts of the neoplasm there was a small amount of squamous epithelium."

CASE XCVI.—*Papillary Growth on the Posterior Wall of the Larynx ; Treatment by Evulsion ; Cure.*

Sarah Ann S., æt. 27, a sewing-machine worker, came under my notice at the Hospital for Diseases of the Throat, March 14th, 1870. She stated that she had suffered from hoarseness for the last two years, and from great soreness in the throat for the last twelve months. Her voice was sometimes phonetic, though always now harsh and unpleasant, and occasionally it subsided into the merest whisper. During the last nine months she had once or twice spat about a teaspoonful of blood. On examination, a large indurated specific ulcer was observed on the under surface of the tongue. There was no appearance of disease in the pharynx, but in the larynx was seen a pink growth with dentated edges, springing erect from the outer arytenoid fold. The appearance was somewhat similar to a small cock's comb (Fig. 83). The right vocal cord was slightly irregular, and the larynx generally was congested. Examination of the chest gave no signs of pulmonary disease.

The patient was treated for some weeks for the congestion by inhalations, and by iodide of potassium ; and the ulcer on the tongue was touched twice a week with solid nitrate of silver. It was not till April 25th that I attempted removal of the growth. On this occasion, in the presence of Mr. Keene, Drs. Collum, Stage, and others, I removed the growth. The patient was seen on the 20th May,

Fig. 83.

when the larynx was perfectly clear, and the voice normal. The ulcer on the under surface of the tongue still remained.

On microscopic examination, the neoplasm was found to be of simple papillary structure.

CASE XCVII.—*Fibroma on the Posterior Wall of Larynx : Treatment by Evulsion ; Cure.*

Emma B., æt. 29, a butcher's wife, applied at the Hospital for Diseases of the Throat, March 31st, 1870, complaining that since July in the previous year her voice had always been lost in damp weather, and that at all times she suffered great pain in speaking, and had to make a great effort in order *to sound* her voice. After using her voice for a short time, she experienced a great sense of fatigue. There was no dysphagia.

On laryngoscopic examination, a pale pink, slightly uneven, growth was seen springing from the posterior wall of the larynx (Fig. 84, and Plate III. fig. 10). The patient was not in good general health when she came to the Hospital, so that it was not till the 2nd of May that, in the presence of Dr. Webb, of Cincinnati,

26

U.S.A., and Dr. Bowditch, I attempted removal of the growth. This was easily effected in two operations, with common laryngeal forceps.

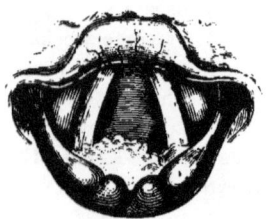

Fig. 84.

On microscopical examination, the neoplasm was found to consist of a compact fibrous structure, the fibres interlacing in every direction. The whole growth was enclosed in an epithelial investment, consisting of about seven layers of cells.

CASE XCVIII.—*Congenital (?) Papillary Growths in the Larynx : Tracheotomy ; Death.*

Thomas M., æt. 2 years and 4 months, the child of a lighterman, was brought to the Hospital for Diseases of the Throat, May 7th, 1870, by his parents, who gave the following account of him :—

From the first moment of birth he had never cried like other babies, but his voice had always been " croupy." For the last six or eight months his breathing had been gradually getting worse, but he had never breathed quite freely since birth. He could not lie down, but was always put to bed with his head very high ; he was very restless in his movements during sleep, and woke frequently. The child was a fine rosy boy, well nourished, and full grown. On trying to make him talk, it was observed that the voice was entirely suppressed, except occasionally, if he was excited, when a slight gruff vocal sound was emitted. His breathing was very noisy and stertorous. There was, however, but slight cough, and no expectoration. His father and mother were both healthy. They had one other child, eight months old, whose voice was quite lusty.

A laryngoscopic examination in this case was attempted, but

with purely negative result, the child being quite unmanageable ; I, however, had no hesitation in expressing opinion that the case was one of congenital growth in the larynx. The complete aphonia and difficulty of breathing seemed to indicate that the growth was of considerable size, and probably in the immediate neighbourhood of the vocal cords. On May 9th the child entered the hospital, and his respiration becoming every hour more embarrassed, tracheotomy was performed on the 10th. The tube was inserted very quickly, with but little hæmorrhage. The child seemed to rally for a time, swallowed stimulants and nourishment, and took notice of his father, whom he would not allow to leave him. Eight hours after the operation, however, the respiration was observed to become very feeble, though there was no stridor or dyspnœa. He sank 12 hours after the operation, and it was observed that he had never coughed nor made any effort to expel mucus from the tube.

On *post-mortem* examination, the diagnosis was at once verified, the glottis being almost completely blocked up with growth. Both vocal cords, the right ventricular band, and half of the left band were entirely covered by a cauliflower mass of warty growths (Plate V. fig. 1), which, on microscopic examination, were found to be of a simple epithelial character (Plate I. fig. 2).—*British Medical Journal,* 1870.

CASE XCIX.—*Fibro-Cellular and Myxomatous Growths on both Vocal Cords ; Treatment by Evulsion ; Cure.*

Mr. L. T., æt. 27, the son of a medical practitioner at Devonport, consulted me, May 18th, 1870, on account of hoarseness, which had existed for two years. The patient was in excellent health, and there was no other symptom of disease. His voice, when I saw him, was distinctly phonetic, but he said that it was always shrill or hoarse, and that he had never been able to sing. The breathing was not in the least embarrassed, even after running or other active exertion.

On laryngoscopic examination, a very large, pink, lobulated, growth was seen occupying the anterior three-fourths of the right vocal cord, and in deep inspiration projecting across the left vocal cord. The portion of the growth furthest from its attachment was brighter in colour and quite transparent. There appeared also to be a

small wart on the left cord (Fig. 85, and Plate III. fig. 11).
The glottis, with the exception of the posterior sixth, was entirely

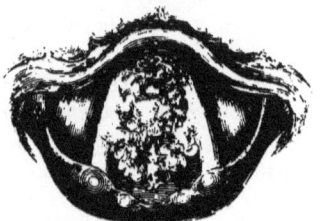

Fig. 85.

occluded.   The larynx being very capacious, the growth large,
and the patient steady, several large pieces were easily removed with common laryngeal forceps.   On May 30th the growth
on the right vocal cord was reduced to about a fourth of its former
dimensions (Plate III. fig. 12), and it was now apparent that the
small growth on the left side was attached to the under surface of the
cord.   Considerable difficulty was experienced in seizing this latter
growth, but it was ultimately removed with antero-posterior laryngeal
forceps.   The remaining portions of the growth on the right vocal
cord were subsequently eradicated, and early in November the voice

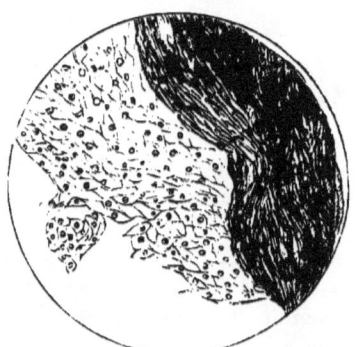

Fig. 86.

was perfectly restored.   The portions removed were of very varied
character and consistence.   Some pieces were hard, others were
of medium consistence, and one fragment, about the size of a large
currant, was gelatinous and semi-transparent.

The following is the report of a microscopical examination, made by my brother, Mr. Stephen Mackenzie :—"Specimens submitted for examination exhibited various kinds of structure in different parts. In some portions the connective tissue predominated, whilst in others the fibrous arrangement was exceedingly regular and compact. In this part many of the fibres were elongated and fusiform, and the structure was almost that of a fasciculated sarcoma. One portion of it was almost gelatinous, and consisted of a nearly homogeneous intercellular substance and a few irregular caudate cells. In some parts there were an abundance of papillæ. The growth much resembled, in the first portion described, the connective-tissue tumour of Vogel, while in the softer portions it approximated closely to a myxoma." (Fig. 86, and Plate I. fig. 4.)

CASE C.—*Papillomatous Growth on the Left Vocal Cord; Treatment by Evulsion; Cure.*

John F., æt. 51, a traveller, applied at the Hospital for Diseases of the Throat, May 23rd, 1870, on account of hoarseness, from which he had suffered for five or six months. He attributed the dysphonia to frequent exposure to cold and variable weather. There was no other symptom of any kind.

On examination with the laryngoscope, a large, round, slightly mammillated, growth, of the colour of the surrounding mucous membrane, was seen to occupy the anterior two-thirds of the left vocal cord.

On May 26th the growth was entirely removed with common forceps, in the presence of Mr. Julius Sankey and Mr. Pugin Thornton. The patient at once spoke clearly, and a fortnight later his voice was quite normal and the larynx free from any appearance of growth.

On microscopic examination, the growth was found to be of papillary nature, the papillæ being very abundant, but unusually short. The central portions of the tumour contained rather more connective tissue than is usual in papillomata.

APPENDIX B.

— o —

SHORT REPORTS OF A FEW CASES IN WHICH
RADICAL TREATMENT WAS NOT ADOPTED.

APPENDIX B.

——o——

CASE I.—*Aphonia of twenty years' duration from Growths on both Vocal Cords ; no Treatment.*

Elijah S., æt. 53, an inspector of police from Tunstall, consulted me, November 17th, 1863, on account of loss of voice, which had existed for twenty years.

On laryngoscopic examination, the right vocal cord was seen to be entirely occupied by a flap-like growth, and on the left vocal cord there were also two small mammillary neoplasms. The patient attributed the aphonia to severe and repeated ulceration of the throat. He had travelled a long distance for an opinion, but was unable to remain in town for curative treatment.

CASE II.—*Small Growth on the Posterior Wall of the Larynx ; no increase after interval of two years ; no Radical Treatment recommended.*

Mrs. B. was first seen by me at the London Hospital in November, 1863 ; she was at that time a patient under the care of Mr. Adams, on account of an abscess in the knee. She had entirely lost her voice for two years, and on examination with the laryngoscope I discovered a small irregular warty growth on the posterior wall of the larynx. It was attached rather to the right side of the median line. Palliative treatment, consisting of inhalations, was used with some benefit. I did not see the patient again till December 6th, 1865. A drawing having been made on the first occasion of her consulting me, I was enabled to make an exact comparison of her larynx with the condition manifested two years previously. There was not the slightest alteration in the size of the growth, and therefore, as her voice was in some measure restored, and as there was no impairment of the respiration, I did not make any attempts at radical treatment.

27

CASE III.—*Congenital* (?) *Aphonia; Excrescences in the Larynx; Death.*

The patient, a child, aged between 3 and 4 years, was brought to the Hospital for Diseases of the Throat early in 1864. She could not be examined with the laryngoscope, principally on account of the shape and position of the epiglottis. The case was a very interesting one, as the aphonia had been congenital, or had occurred immediately after birth. The mother had never known the child to utter a vocal sound. When it cried, tears came into its eyes, but it produced no sound.

On introducing the finger into the larynx, during life, a hard round tumour was felt beneath the epiglottis. It seemed of cartilaginous or bony character, and was believed to be a growth connected with the thyroid cartilage. There appeared to be no doubt as to the nature of the case, and the patient was exhibited at the Medical Society of London, where several hospital physicians and surgeons carefully examined the larynx, and agreed as to the nature of the disease. The child subsequently died in an epileptic fit.

Instead of a cartilaginous tumour, warty growths were discovered in the larynx. Some small warts, the largest of which was about the size of a tare, were found on the left vocal cord, and beneath the vocal cords on each side. The mucous membrane generally was of a warty or granular character.

In this case, it will be seen that a correct diagnosis was arrived at, although one step in the investigation gave fallacious results. The digital examination indicated the presence of a hard growth at the base of the epiglottis, but, on *post-mortem* examination, soft warty excrescences were found on the left vocal cord. It became interesting to ascertain how this discrepancy in diagnosis had arisen, and on consideration, there could be no doubt that it had been caused by the hyoid bone being pressed down, when the finger was introduced into the mouth, and by its having been felt through the epiglottis. Though such cases are rare, and though the mistake could only happen before the cornua of the hyoid bone are developed, this case is very instructive in a practical point of view, more especially as, previous to the publication of this case, no attention has been called to this source of fallacy.—(*Transactions of the Pathological Society*, vol. xvi. pages 37 and 38.)

CASE IV.—*Large Growth on the Right Vocal Cord ; Treatment declined ; Death by Suffocation.*

Mr. D. B., æt. 42, a merchant, consulted me on the 5th of July, 1865, by the advice of Dr. Scofield, of Birmingham, on account of hoarseness and occasional attacks of dyspnœa, which had existed for some months.

On examination with the laryngoscope, I discovered an irregular, coarsely lobulated growth, springing from the centre of the right vocal cord, and obscuring more than four-fifths of the cavity of the glottis. I recommended the patient to remain in town and have the growth removed, but this he was unwilling to do ; I therefore wrote to Dr. Scofield, advising that, in the event of his symptoms becoming more urgent, tracheotomy should be performed. In this advice I was also supported by Mr. Furneaux Jordan, who saw the patient on his return home. The patient, however, declined an operation, and he died suddenly, August 1st, before the windpipe could be opened.

CASE V.—*Out-growths on the Vocal Cords ; no Mechanical Treatment recommended.*

Mr. B., æt. 45, was brought to me, June 28th, 1866, by Dr. Hyde Salter, on account of hoarseness, which had been coming on gradually for two years. This was the only symptom.

On examination with the laryngoscope, a sessile nodule was seen on each vocal cord, the one on the right side being the larger (Fig. 87).

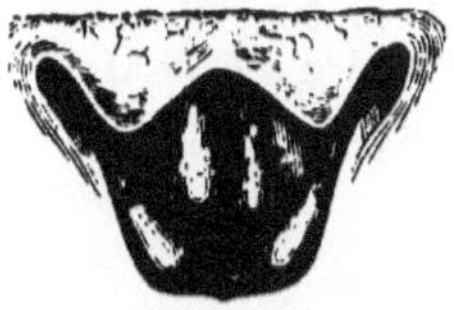

Fig. 87.

As these small projections appeared to be thoroughly incorporated in the subjacent tissues, I merely recommended the application of mineral astringents. (*Medical Times and Gazette*, June 13th, 1868.)

CASE VI.—*Growth on the Left Vocal Cord ; no Treatment recommended.*

The Rev. J. H., æt. 73, living in Wales, was. brought to me by Mr. Richard James, of Notting Hill, Sept. 16th, 1866. The patient's voice had been impaired for between three and four years, but had been completely suppressed for two years.

On laryngoscopic examination, a small nodal excrescence was seen on the centre of the left vocal cord. There was no distinct line of demarcation between the growth and the vocal cord, and as the patient was provided for by a pension, I did not think it desirable, at his advanced age, to recommend any operative procedure.

CASE VII.—*Growth on the Posterior Wall of Larynx; no Treatment.*

Miss I., æt. 49, first consulted me October 15th, 1866, on account of hoarseness and shortness of breath. The former symptom had existed for between three and four years, but the dyspnœa, which was of a stridulous character, had only been occasionally present during the last seven months.

On making a laryngoscopic examination, a pyramidal growth, about the size of a small kidney-bean, was seen projecting from the posterior wall of the larynx, upwards and forwards, into the cavity of the larynx, above the glottis. The patient came up from Devonshire, where she had been residing, on account of a supposed delicacy of the lungs ; but a careful stethoscopic examination failed to elicit any evidence of phthisis. Miss I. was very nervous, and refused to allow any attempt at removal. She subsequently placed herself under the care of my friend Surgeon-Major Wyatt, who has been able to give her some relief by the application of mineral astringents to the larynx.

CASE VIII.— *Very small Growth on the inter-Arytenoid Fold ; no Mechanical Treatment recommended.*

Dr. A., a professor of languages, consulted me, October 18th, 1866, on account of hoarseness and cough. These symptoms had only existed for a few weeks, but he had long been subject to bronchitis in the winter. This gentleman was rather nervous, and it would have been very difficult to adopt operative procedure in his case, more especially as the growth was so very small. By the application of caustic solutions and stimulating inhalations, and with rest of the vocal organs, the symptoms gradually diminished, but there was no alteration in the size of the growth.

CASE IX.—*Congenital Growth ; Tracheotomy recommended but not permitted ; Death by Asthenia ; Post-mortem Examination.*

Sophia W., æt. 2 years and 11 months, was sent to me by Dr. Sydney Ringer, July 9th, 1867.

The patient was very weak and much emaciated, and her mother stated that from the time of her birth it was noticed that she never cried aloud, although she was able to make a harsh kind of crowing noise. As a baby, she always had great difficulty in sucking, and was obliged to stop very often to fetch her breath. The mother was unwilling to allow tracheotomy to be performed, and a few days after I first saw her, the child gradually sank from asthenia. On examining the larynx after death, the vocal cords were found to be covered by a fringe of warty growths.

CASE X.—*Small Excrescence on the Right Vocal Cord ; Mechanical Treatment not recommended.*

Mr. John T., æt. 61, residing at Bolton, consulted me, September 3rd, 1867, on account of general debility, loss of appetite, and hoarseness. He said that he had not felt well for some time, but that the aphonia had only existed for eight months, and he stated also that his voice was certainly not worse now than it had been three months previously, though it was always rendered more hoarse by cold and damp. He had no cough nor expectoration. Until the last year or so he had always enjoyed good health.

On laryngoscopic examination, I discovered a small mammillated growth, the size of half a pea, springing from the centre of the right vocal cord, and preventing the cords from approximating closely in phonation. The symptoms in this case were not, in my opinion, sufficiently urgent to warrant immediate mechanical interference. I therefore recommended Mr. T. to spend the winter in a warm climate, and to return to me in the spring, when, if any important vocal symptoms had arisen, I proposed to remove it. This patient did not apply to me again ; but I heard from Dr. Livy, of Bolton, his usual medical attendant, that his voice continued in the same state.

CASE XI.—*Growths on each Vocal Cord ; no Treatment.*

Mrs. B., æt. 52, consulted me in the spring of 1868, by the advice of Dr. Holman, of Hampstead.   At that time she was suffering from enlargement of the veins of the larynx and posterior nares.   Her complexion was pale, but the superficial capillaries were enormously enlarged and varicose.   She constantly suffered from hæmorrhage from the throat and posterior nares.   I occasionally saw this lady, and her condition did not vary much ; but in May, 1870, she came to me on account of hoarseness.   I observed a small pea-like growth situated on the cartilaginous portion of each vocal cord.   In view of the great disposition to hæmorrhage, I felt some hesitation in operating on the growths, and I accordingly proposed a consultation with an eminent surgeon.   The patient, however, declined treatment.

CASE XII.—*Sessile Growth on the Left Vocal Cord ; Treatment declined.*

Captain G., æt. 42, consulted me on the 12th of March, 1869, on account of hoarseness, which had existed for some months.   The patient had resided for many years in Australia, and had been much exposed to night air.   His general health was not at all affected. On examination with the laryngoscope, I discovered a sessile, red, growth on the left cord,. which prevented approximation in phonation.

I cautioned the patient that it was possible that more serious symptoms might in time arise ; but as at the time of his consulting me there was no other discomfort than the hoarseness, he objected to undergo any operative treatment.

APPENDIX C.

TABLE OF ONE HUNDRED CONSECUTIVE CASES TREATED
BY THE AUTHOR.

—— σ ——

₊ Where the nature of the Growth was inferred from its naked-eye appearances, and no microscopic examination was made, the " Pathological Nature " is enclosed in parentheses.

Where "antero-posterior" forceps are stated to have been used, it means that this variety of the author's common laryngeal forceps has been employed.

28

No. of Case.	Date.	Sex.	Age.	Occupation.	Symptoms.	Situation.	Pathological Nature.
1	June, 1862	M.	40	Bookseller	Aphonia	Epiglottis	(Papilloma)
2	April, 1863	F.	42	Housewife	Aphonia ...	Right Ary-epiglottic Fold and Ventricular Band.	(Papilloma)
3	April, 1863	M.	44	Carpenter...	Aphonia and Dyspnœa.	Almost the entire lining membrane of Larynx.	Papilloma...
4	April, 1863	F.	35	Mechanic's Wife.	Aphonia ...	Both Vocal Cords, and below Vocal Cords.	(Papilloma)
5	May, 1863	M.	45	Gas-fitter ...	Aphonia	Left Vocal Cord ...	Papilloma...
6	May, 1863	M.	40	Waiter ...	Dysphonia and frequent Cough.	Both Vocal Cords ...	(Papilloma)
7	Aug. 1863	M.	41	Shoemaker	Dysphonia ...	Anterior Commissure of Vocal Cords.	Fibroma ...
8	Jan. 24, 1864	M.	26	Sailor ...	Dysphonia and Dysphagia.	Epiglottis and Right Ventricular Band.	Benign Epithelial Growth.
9	Jan. 1864	F.	45	Lady ...	Aphonia and great Dyspnœa.	Right Vocal Cord ...	Papilloma...
10	April, 1864	F.	30	Milliner ...	Aphonia	Both Vocal Cords ...	(Papilloma)
11	Oct. 1864	M.	4	Mechanic's child.	Aphonia, stridulous Breathing, and attacks of Suffocation.	Just above Anterior Commissure of Vocal Cords, and beneath Vocal Cords.	Papilloma...
12	Nov. 1864	F.	6	Labourer's child.	Aphonia	Both Vocal Cords ...	Papilloma...
13	Jan. 1868	F.	28	Fish-hawker	Constant tickling in Larynx (neither Aphonia nor Dysphonia).	Posterior part of right Vocal Cord.	(Papilloma)
14	Jan. 1865	M.	46	Hawker ...	Aphonia	Right Ventricular Band and both Vocal Cords.	Benign Epithelial Growth.

Treatment.	Result.	References and Remarks.
Destruction with caustic solutions.	Great improvement.	Seen by Professor Czermak. (*The Use of the Laryngoscope*, second edition, p. 86.) Voice still hoarse. Slight diminution in size of Growth.
Destruction with caustic solutions.	Improvement ...	Patient was obliged to return home suddenly. She left with a distinctly phonetic but rather harsh voice. (*The Use of the Laryngoscope*, second edition, p. 87.)
Evulsion with tube-forceps, and application of escharotics.	Cure	Seen by Drs. Czermak, Frodsham, and others. (*The Use of the Laryngoscope*, second edition, p. 121.) Recurrence on both ventricular bands, March 7, 1870. Second series of four operations with common laryngeal forceps. Cure.
Evulsion with tube-forceps ...	Cure ...	Sent by Mr. Brown, of Finsbury. (*The Use of the Laryngoscope*, second edition, p. 123.)
Evulsion with tube-forceps ...	Cure ...	Seen by Drs. Czermak, Wahltuch, and others. (*The Use of the Laryngoscope*, second edition, p. 125.)
Evulsion with tube-forceps, and incision of base.	Cure ...	Seen by Dr. George Johnson, &c. (*The Use of the Laryngoscope*, second edition p. 126.)
Evulsion with common laryngeal forceps.	Cure	Dr. George Johnson and Mr. Francis Mason present at operation. (*The Use of the Laryngoscope*, second edition, p. 128.)
Evulsion with common laryngeal forceps.	Cure ...	Seen by Mr. George Evans. Fresh Growth (Feb. 1866) beneath anterior Commissure of Vocal Cords. Removal with tube-forceps, and Cure. Seen on second occasion by Drs. Tatham and Thurgar.
Partial evulsion with tube-forceps.	Great improvement.	Breathing became more natural. Voice remained slightly hoarse. Portions removed were examined by Dr. Andrew Clark.
Evulsion with common laryngeal forceps.	Cure ...	Sent by Mr. Parsons, of Bridgewater. (*The Use of the Laryngoscope*, second edition, p. 130.)
Evulsion with tube-forceps ...	Cure ...	(*Transactions of Pathological Society*, vol. xvi. p. 38.)
Evulsion with tube-forceps ...	Cure ...	*Ibid.*, p. 39.
Evulsion with tube-forceps ...	Cure	Seen by Dr. Mill Frodsham.
Partial evulsion with common laryngeal forceps.	Improvement ..	Patient discontinued attendance. Microscopical examination by Dr. Andrew Clark.

No. of Case.	Date.	Sex.	Age.	Occupation.	Symptoms.	Situation.	Pathological Nature.
15	Jan. 1865	M.	12	... ...	Aphonia and serious Dyspnœa.	Right Vocal Cord and beneath anterior Commissure.	Papilloma ..
16	May 10, 1865	M.	37	Vocalist ...	Dysphonia ...	Posterior Wall of Larynx.	Fibroma ...
17	June, 1865	F.	31	Painter's wife.	Aphonia	Left Vocal Cord	Benign Epithelial Growth.
18	July 5, 1865	M.	47	Engine-driver.	Dysphonia ...	Right Vocal Cord ...	Papilloma...
19	July, 1865	M.	34	Mechanic ...	Dysphonia ...	Just above anterior insertion of Vocal Cords, attached to Epiglottis.	(Fibroma)...
20	Sept. 16, 1865	M.	35	Labourer ...	Aphonia	Both Vocal Cords ...	(Papilloma)
21	Dec. 1865	M.	36	Tailor ...	Great Dyspnœa, Dysphonia, and Cough.	Right Vocal Cord ...	(Papilloma)
22	Jan. 18, 1866	F.	45	Messenger's wife.	Aphonia and serious Dyspnœa, with threatened Suffocation.	Left Vocal Cord and Anterior Commissure of Vocal Cords.	Papilloma...
23	Jan. 1866	M.	34	Coachman	Aphonia	Both Vocal Cords and Ventricular Band.	(Papilloma)
24	Feb. 1866	F.	24	Housemaid	Severe Dysphonia and occasional attacks of Dyspnœa.	Beneath anterior Commissure of Vocal Cords.	(Papilloma)
25	Feb. 1866	F.	44	Housewife	Dysphagia ...	Upper Surface of Epiglottis.	(Cystic Growth.)
26	Mar. 1866	M.	35	Vocalist ...	Aphonia	Anterior Inter-arytenoid fold.	Papilloma...
27	Mar. 1866	F.	57	Stonemason's wife.	Dysphonia ...	Right Vocal Cord ..	(Fibroma)...
28	April, 1866	F.	45	Mechanic's wife.	Odynphagia, Dysphonia, and Dyspnœa.	Right Ventricular Band.	Fibroma ...
29	June, 1866	M.	60	Gentleman	Aphonia	Both Vocal Cords ...	Papilloma...
30	July, 1866	M.	49	Stableman	Dysphonia ...	Both Ventricular Bands.	(Papilloma)

Treatment.	Result.	References and Remarks.
Evulsion with tube-forceps ...	Cure ...	Seen by Mr. Mason. Portion removed was examined by Dr. Andrew Clark, and exhibited at the Pathological Society. (*The Use of the Laryngoscope*, second edition, p. 133.)
Evulsion with tube-forceps ...	Cure	Speaking voice restored, but no power in singing. Seen by Drs. Frodsham and Dale.
Evulsion with tube- and common laryngeal forceps.	Cure	Seen by Dr. Pratt. Microscopical examination by Dr. Andrew Clark. (*The Use of the Laryngoscope*, second edition, p. 134.)
Evulsion with tube-forceps ...	Cure	Patient ceased attendance before all the Growth was removed. Case seen by Mr. Evans.
Two unsuccessful attempts at removal with tube-forceps.	Negative	Patient came from Leeds, and returned the same day.
Evulsion with tube-forceps ..	Great improvement.	Patient regained gruff but distinctly phonetic and satisfactory voice. Seen by Drs. Baxter and Clark, of Melbourne.
Evulsion with antero-posterior laryngeal forceps.	Cure ..	Case seen by Drs. Pratt, Tatham, and Lanchester.
Partial evulsion with wire loop	Great improvement.	Considerable relief in breathing, and almost complete restoration of voice. Portions removed, with microscopic specimens, exhibited at the Pathological Society, Feb. 1866. (*Transactions*, vol. xviii. p. 33.)
Evulsion with tube-forceps ...	Slight improvement.	Discontinued attendance after a few sittings, at each of which pieces were removed. A portion remained on right Ventricular Band and on left Vocal Cord.
Evulsion with tube-forceps ...	Cure ...	In this case, although a very small portion was left, the symptoms entirely disappeared. There has been no recurrence. Case seen by Drs. Tatham and Truell.
Free incision and application of solid nitrate of silver. ...	Cure	Sent by Mr. Gayton. (*Medical Times and Gazette*, June 13, 1868.)
Evulsion with tube-forceps ...	Cure	Dr. Pratt present.
Incision of base of Growth with laryngeal lancet.	Cure	Dr. Tatham and Dr. Chisholm, of Charleston, U.S.A., present.
Evulsion with common laryngeal forceps.	Cure	Sent by Mr. Hind, of Gravesend. Fragment examined by Dr. Andrew Clark. (*Medical Times and Gazette*, June 13, 1868.)
Evulsion with tube-forceps ...	Cure ...	Brought by Mr. Paget. (*Medical Times and Gazette*, June 13, 1868.)
Evulsion with common laryngeal forceps.	Cure ...	Seen by Mr. Evans before and after treatment.

No. of Case.	Date.	Sex.	Age.	Occupation.	Symptoms.	Situation.	Pathological Nature.
31	Nov. 1866	M.	23	Vocalist ...	Dysphonia and frequent Cough.	Left Vocal Cord	(Papilloma)
32	Dec. 1866	F.	50	Laundress...	Dyspnœa ...	Left Capitulum Santorini.	(Papilloma)
33	Dec. 17, 1866	M.	37	Vocalist ...	Dysphonia ...	Right Vocal Cord ...	(Papilloma)
34	Dec. 1866	F.	46	Lady	Aphonia ...	Right Vocal Cord ...	(Papilloma)
35	Jan. 13, 1867	M.	42	Railway Porter.	Dysphonia ...	Right Vocal Cord ...	(Papilloma)
36	Jan. 17, 1867	M.	54	Gardener ...	Aphonia and serious Dyspnœa.	Right Vocal Cord ...	Small (fibrous) Growths. intimately associated with subjacent parts.
37	Feb. 1867	M.	40	Overseer in coal-mine.	Aphonia ...	Right Vocal Cord and under surface of Epiglottis.	Papilloma...
38	Mar. 1867	F.	56	Lady ...	Dysphonia ...	Posterior wall of Larynx.	(Papilloma)
39	Mar. 1867	M.	27	Vocalist ...	Dysphonia ...	Left Vocal Cord ...	Fibroma ...
40	Mar. 1867	M.	32	Labourer ...	Complete Aphonia	Symmetrical Growths on both Vocal Cords.	(Papilloma)
41	Mar. 1867	F.	28	Servant ...	Slight Dysphonia...	Under surface of Epiglottis.	Fibro-cellular Growth.
42	May, 1867	M.	52	Missionary	Aphonia	Right Vocal Cord ...	(Fibroma)...
43	May, 1867	F.	53	Housewife	Aphonia, Dyspnœa, and Cough.	Posterior Wall of Larynx.	Papilloma...
44	June, 1867	F.	41	Housewife	Dysphonia, slight Cough, Dyspnœa, and occasional feeling of Suffocation.	Right Ventricular Band.	Papilloma...
45	June 13, 1867	F.	30	Watch-maker's wife.	Aphonia and great Dyspnœa.	Both Vocal Cords ...	(Papilloma)
46	July 3, 1867	F.	31	Lady	Aphonia ...	Whole lining membrane of Larynx.	(Papilloma)

Treatment.	Result.	References and Remarks.
Incision of base of Growth ...	Cure ...	The Growth atrophied after incision of base. Operated on in presence of Dr. Taylor, Dr. Merryweather, and others.
Evulsion with tube-forceps ...	Cure ...	Patient seen by Mr. Evans.
Evulsion with tube-forceps ...	Cure	Case seen in consultation with Mr. Evans.
Division of base of Growth ...	Cure	Case seen before and after treatment by Drs. Hun and Lockwood, of New York. (*Medical Times and Gazette*, June 13, 1868.)
Evulsion with tube- and antero-posterior forceps.	Cure ...	Case seen during treatment by Dr. Pogojeff, of Odessa, and Mr. Du Pasquier.
Attempted evulsion with common laryngeal forceps and abscission. Tracheotomy.	Improvement in respiration.	Tracheotomy performed by Mr. Evans. Difficulty of breathing relieved, but Aphonia remained. Canula always worn.
Evulsion of all but a minute piece with tube-forceps.	Cure ...	Sent by Dr. Griffiths, of Swansea. Recurrence May 25, 1868, in same situations. Removal; no further recurrence. (*Medical Times and Gazette*, June 13, 1864.)
Incomplete destruction by galvanic cautery.	Negative ...	Case sent by Dr. Addington Symonds.
Evulsion with tube-forceps ...	Cure ...	Case sent by Dr. Marion Sims.
Evulsion with antero-posterior laryngeal forceps.	Great improvement.	Patient satisfied with phonetic but hoarse voice. Dr. McCall Anderson present at operation.
Destruction by galvanic cautery.	Cure ..	Case seen before treatment by Drs. Atkinson and Macaldin.
Destruction by galvanic cautery.	Cure	Inflammation of the Larynx followed. Mr. Lennox Browne present at the operation.
Evulsion with tube-forceps ...	Cure ...	Case seen frequently by Dr. Peléchin, of St. Petersburg.
Evulsion with tube-forceps ...	Cure ...	May, 1868. Larynx remained quite healthy. Tumour examined microscopically by Dr. Andrew Clark.
Partial evulsion with tube-forceps.	Improvement ..	Patient was advanced in pregnancy when treatment commenced, and was unable to remain in Hospital on that account. Tracheotomy was subsequently performed by Mr. De Berdt Hovell, of Clapton.
Evulsion with tube-forceps ...	Cure ...	Case known to Dr. Money, of Brighton. Recurrence June 27, 1870. Second series of operations with tube-forceps, and recovery of voice.

No. of Case.	Date.	Sex.	Age.	Occupation.	Symptoms.	Situation.	Pathological Nature.
47	July 9, 1867	M.	45	Shoemaker	Aphonia and Cough	Both Vocal Cords ...	(Papilloma)
48	July, 1867	M.	30	Furniture-dealer.	Aphonia ...	Under surface of Epiglottis.	Fibro-cellular Growth.
49	July, 1867	F.	53	Bible-reader	Aphonia, Dyspnœa, and severe Dysphagia.	Under surface of Epiglottis and anterior Commissure of Vocal Cords.	Fasciculated Sarcoma.
50	Sept. 10, 1867	M.	64	Merchant ...	Aphonia	Right Vocal Cord ...	(Papilloma)
51	Oct. 1867	M.	55	Dispenser ...	Aphonia ...	Right Vocal Cord ...	(Papilloma)
52	Oct. 1867	F.	65	Lady ...	Aphonia and Dyspnœa.	Right Vocal Cord ...	Fibro-cellular Growth.
53	Nov. 1867	M.	26	Butcher ...	Aphonia	Both Vocal Cords ...	(Papilloma) Great general Hyperæmia.
54	Dec. 12, 1867	M.	43	Farmer ...	Aphonia	Right Vocal Cord ...	Benign Epithelial Growth.
55	Dec. 16, 1867	M.	13	Washerwoman's son.	Dysphonia and dyspnœa.	Left Vocal Cord ...	(Papilloma)
56	Dec. 1867	M.	34	Sugar-baker	Dysphagia ; constant desire to clear throat.	Under-surface of left side of Epiglottis.	Fibroma ...
57	Dec. 1867	M.	17	Engine-fitter	Aphonia and Shortness of Breath, but no Dyspnœa.	Left Vocal Cord and under-surface of Epiglottis.	(Papilloma)
58	Jan. 7, 1868	M.	10	Artisan's child.	Aphonia	Right Ventricular Band.	Benign Epithelial Growth.
59	Jan. 13, 1868	M.	42	Carman ...	Aphonia	Right Vocal Cord and Right Ventricle.	Fasciculated Sarcoma.
60	Feb. 6, 1868	F.	23	Servant ...	Dyspnœa	Right Vocal Cord ...	Papilloma...
61	Feb. 8, 1868	F.	31	Lady ...	Aphonia	Right Vocal Cord ...	(Papilloma)
62	Feb. 1868	F.	57	Fish-hawker	Dysphonia and Cough.	Right Vocal Cord ...	(Papilloma)

Treatment.	Result.	References and Remarks.
Evulsion with common laryngeal forceps and tube-forceps.	Cure ...	Case seen during treatment by Surgeon-Major Trestrail and the Rev. David Bell, M.D.
Evulsion with tube-forceps	Cure	Surgeon-Major Trestrail present at operation.
Partial destruction by electric cautery, and subsequent evulsion with cutting forceps.	Considerable improvement.	Dyspnœa and Dysphagia quite relieved ; voice sufficiently restored to enable her to resume her employment. Microscopical examination by Mr. Stephen Mackenzie.
Evulsion with common laryngeal forceps.	Cure	Operation performed in presence of Mr. Lennox Browne. Case sent by Mr. Ince.
Partial evulsion. Crushing with strong forceps.	Cure	Surgeon-Major Trestrail and others present.
Evulsion with common laryngeal forceps.	Cure	This Growth, the size of a cherry, was removed completely in one operation. (*Transactions of Pathological Society*, vol. xix. p. 60.)
Partial evulsion with antero-posterior forceps.	Improvement ...	Patient discontinued attendance after two pieces of Growth had been removed in the presence of Surgeon-Major Trestrail.
Destruction with caustic and astringent solutions.	Cure	Patient frequently seen by Surgeon-Major Trestrail and others.
Evulsion with tube-forceps ...	Cure	Drs. Welch and Peléchin present.
Evulsion with common laryngeal forceps.	Cure ...	Case seen by Dr. Alexander Fox.
Evulsion of a portion with tube- and common forceps and Stoerk's écraseur.	Cure ...	Sent by Dr. Johnston, of Barnstaple. Seen again in March, 1869, when there was no return of Growth. (*Medical Times and Gazette*, June 13, 1868.)
Evulsion with tube-forceps ...	Cure	Dr. Alexander Hewan and others present during treatment.
Evulsion and crushing with common antero-posterior forceps.	Great improvement. Still under treatment.	Constant recurrence, the tumour appearing to grow afresh between each visit. Patient seen and fragment examined microscopically by Dr. Fenwick and Mr. Stephen Mackenzie.
Evulsion with common laryngeal forceps.	Cure ...	Operated on in presence of Surgeon-Major Trestrail and Dr. Wilkie.
Evulsion with tube- and common laryngeal forceps.	Cure ..	Sent by Mr. Harston, of Islington. Fresh Growth in consequence of not all having been removed on the first occasion. Complete recovery ultimately. (*Medical Times and Gazette*, June 13, 1868.)
Evulsion of fragments with tube-forceps.	Negative	Patient did not apply after second visit.

29

No. of Case.	Date.	Sex.	Age.	Occupation.	Symptoms.	Situation.	Pathological Nature.
63	Feb. 1868	M.	10	Gentleman's son.	Aphonia and Dyspnœa.	Anterior Commissure of Vocal Cords.	Papilloma...
64	April 21, 1868	F.	66	Spinster lady.	Aphonia and severe Dyspnœa.	Both Vocal Cords, nearly filling the Glottis.	(Papilloma)
65	May 4, 1868	M.	34	Twine-spinner.	Aphonia ...	Left Ventricular Band	Papilloma ..
66	June 4, 1868	M.	30	Waiter	Dysphonia	Both Vocal Cords ...	(Papilloma)
67	June, 1868	M.	36	Woolpacker	Aphonia ...	Two Growths on right Vocal Cord.	(Papilloma)
68	July 3, 1868	M.	8	Wine-merchant's son.	Almost complete Aphonia.	Right Vocal Cord ...	Papilloma...
69	July 7, 1868	F.	12	Mechanic's child.	Dysphonia and severe Dyspnœa.	Both Vocal Cords ...	(Papilloma)
70	July 9, 1868	M.	48	Woolpacker	Aphonia ...	Right Vocal Cord ...	(Papilloma)
71	July 13, 1868	M.	27	Telegraphist	Aphonia, slight Dyspnœa, hacking Cough.	On and beneath both Vocal Cords.	(Papilloma)
72	Aug. 1868	M.	47	Hawker ...	Dysphonia ...	Right Vocal Cord ...	(Papilloma)
73	Oct. 1868	F.	18	Artist	Dysphonia and Dyspnœa.	Left Vocal Cord, occupying anterior two-thirds of Glottis.	(Fibro-cellular Growth.)
74	Oct. 1868	M.	41	Labourer ...	Dysphonia and Dyspnœa.	Whole length of left Vocal Cord.	Fibro-epithelial Growth.
75	Nov. 1868	M.	15	Page ...	Dysphonia ...	Left Vocal Cord ...	(Papilloma)
76	Dec. 1868	F.	31	Housewife	Aphonia ..	Above and below Anterior Commissure and Right Vocal Cord.	(Papilloma)

Treatment.	Result.	References and Remarks.
Evulsion with tube- and common forceps.	Cure ...	The Growth, $\frac{2}{3}$ of an inch long and $\frac{1}{2}$ an inch broad, was of not more than six months' standing. Patient sent by Mr. Graves, of Gloucester. (*Medical Times and Gazette*, June 13, 1868.)
Evulsion of a portion with common forceps; subsequently Tracheotomy and Thyrotomy.	Cure	Tracheotomy and Thyrotomy performed by Mr. Couper. Complete restoration of voice. Recurrence 2½ years later.
Evulsion with common laryngeal forceps.	Cure	Case seen by Dr. Carlill, Dr. Sykes, and Dr. Henry Roberts (Manchester).
Evulsion with common laryngeal forceps.	Cure ...	Dr. Simpson, of Manchester, and Mr. Balmanno Squire present.
Partial evulsion with tube-forceps.	Improvement ...	Patient recovered and retained phonetic but gruff voice. Vocal Cords remained rough.
Evulsion with tube-forceps, and destruction with caustic solutions.	Cure ...	Patient sent to me by Mr. Sankey, of Maidstone. Fresh Growth two years later on posterior Wall, and on left Vocal Cord. Repetition of treatment and recovery of good voice.
Tracheotomy and Thyrotomy	Improvement ...	Operation performed by Mr. Evans. Dyspnœa cured, but the aphonia unrelieved.
Evulsion with antero-posterior forceps.	Cure	Drs. Jagielski, O'Keefe, and Chatterton present.
Evulsion with tube and common laryngeal forceps.	Cure ...	The patient came from Malta, and was operated on, on one occasion, in the presence of Dr. Gray, of Oxford.
Evulsion with antero-posterior forceps.	Cure ...	Case first seen by Mr. Lennox Browne.
Evulsion with tube-forceps ...	Cure	Paralysis of abductor of left Vocal Cord coming on two months after evulsion of Growth, Tracheotomy became necessary. Operation by Mr. Evans. Patient died eighteen months later, and at autopsy there was no vestige of the Growth.
Evulsion of a portion with common antero-posterior forceps. Tracheotomy. Excision with cutting forceps.	Cure	Patient ceased attendance, content with improvement. Increase of remnant, and Tracheotomy by Mr. Thornton, October, 1869. Subsequent evulsion of all the Growth with cutting forceps, and removal of tube.
Evulsion with tube-forceps ...	Cure ...	Present, Dr. Greenaway, Dr. Gourlay, and others. Recurrence in same position, September, 1869. Second series of operations in presence of Dr. Stage, of Copenhagen, Mr. James Keene, &c. Cure.
Evulsion with tube-forceps	Cure	Case sent by Dr. Bäumler, by whom the patient was examined with laryngoscope before and after treatment.

No. of Case.	Date.	Sex.	Age.	Occupation.	Symptoms.	Situation.	Pathological Nature.
77	Jan. 1869	M.	27	Vocalist ...	Cough, slight Dysphonia in ordinary voice, total loss of singing voice.	Posterior Wall of Larynx.	Papilloma...
78	Feb. 1869	M.	42	Publican ...	Aphonia and Dyspnœa.	Left Vocal Cord ...	Fibroma ...
79	Mar. 1869	M.	31	Stonemason	Dysphonia and Dyspnœa.	Beneath Anterior Commissure of Vocal Cords.	Adenoma ...
80	April 21, 1869	M.	60	Merchant ...	Hoarseness and troublesome Cough.	Both Vocal Cords ...	(Papilloma)
81	May 27, 1869	F.	51	Charwoman	Stridulous Breathing and Dyspnœa.	Posterior Wall of Larynx beneath the Vocal Cords.	(Papilloma)
82	May 14, 1869	F.	21	Servant ...	Never able to sound her voice; great Dyspnœa one year.	Both Vocal Cords ...	Papilloma...
83	May, 1869	F.	51	Housewife	Dysphonia and Dysphagia.	Right Vocal Cord and under surface of Epiglottis.	Papilloma...
84	June, 1869	F.	21	Servant ...	Great Dyspnœa, attacks of strangulation, occasional Aphonia.	Both Vocal Cords ...	Papilloma...
85	June 28, 1869	M.	22	Farrier ...	Dysphagia and slight hoarseness.	Large globular Growth, occupying right side of Epiglottis.	(Cystic Growth.)
86	June, 1869	M.	43	Farmer ...	Complete Aphonia; severe Dyspnœa.	Just below Anterior Commissure.	(Papilloma)
87	Sept. 18, 1869	M.	47	Silversmith	Severe paroxysmal Cough, Dysphonia, and Dyspnœa.	Right Vocal Cord ...	Epithelioma.
88	Oct. 12, 1869	M.	51	Officer in H. M. S.	Cough and Dysphagia. Voice normal.	Under Surface of Epiglottis.	Adenoma ...
89	Dec. 1869	M.	35	Captain in H. M. Marines.	Uneasy tickling sensation in the throat.	Hyoid Fossa ...	Vascular tumour.

Treatment.	Result.	References and Remarks.
Evulsion with tube-forceps ...	Cure	In September, 1869, Larynx remained healthy. Portion of Growth examined by Dr. Fenwick.
Evulsion with ordinary laryngeal and tube-forceps.	Cure	Seen with Dr. Weber, February 12th, 1869.
Evulsion with tube and ordinary antero-posterior forceps, and Stoerk's écraseur. .	Cure	... Seen by the Rev. Dr. Bell, M.D.
Evulsion with tube-forceps ...	Cure	... Treated, in conjunction with Dr. Simpson, of Manchester.
Laryngotomy. Growths afterwards removed through canula opening in crico-thyroid space.	Cure	... Growth not seen, on account of paralysis of abductors of the Vocal Cords, till ten days after operation. Voice and breathing now normal. Tube has long been removed. Seen by Dr. Boddaert, of Ghent, and Mr. Wordsworth.
Evulsion with tube-forceps and cutting forceps.	Cure	... Case briefly reported in *Lancet*, June 5, 1869, p. 779. The Larynx in this case is not larger than that of a child aged eight years.
Evulsion of Growth under Epiglottis, and portion of that on Vocal Cord, with tube-forceps.	Improvement	Patient discontinued attendance. Growth examined by Dr. Fenwick.
Evulsion with tube-forceps, Stoerk's écraseur, and common laryngeal forceps.	Cure	Patient seen by Dr. Elsberg, of New York.
Incision, evacuation of contents, and application of solid caustic.	Cure	... At the expiration of a month there was not even a scar. Case seen by Dr. Gore Ring.
Evulsion of all but a minute piece with tube-forceps and Stoerk's écraseur.	Great improvement.	Sent by Dr. Lewis, of Basingstoke.
Tracheotomy and Thyrotomy	Death 8 months after operation.	Operation performed by Mr. Wordsworth. Post-mortem examination showed that the whole larynx and tracheal opening, as well as the skin surrounding the tube, were occupied by an enormous mass of warty Growths, all developed since the operation.
Tracheotomy. Subsequent removal of Growth through upper orifice of larynx, with guarded wheel-écraseur.	Improvement ; still under observation.	Tracheotomy by Mr. James Adams. Subsequent removal of Growth in presence of Mr. Clover. (*Transactions of Pathological Society*, vol. xxi. p. 51.)
Evulsion with tube-forceps and excision with cutting forceps.	Cure	... Sent by Dr. Smyly, of Dublin.

No. of Case.	Date	Sex.	Age.	Occupation.	Symptoms.	Situation.	Pathological Nature.
90	Jan. 20, 1870	M.	44	Labourer	Dysphonia; sense of oppression about throat.	Above and below Anterior Commissure.	Papilloma...
91	Jan. 1870	M.	47	Wine-bottler	Aphonia ...	Both Vocal Cords ...	(Papilloma)
92	Feb. 3, 1870	F.	48	Haberdasher	Aphonia 24 years...	Left Vocal Cord, and beneath it.	Benign Epithelial Growth.
93	Mar. 3, 1870	M.	30	Engine-driver.	Dysphonia ...	Left Vocal Cord ...	(Papilloma)
94	Mar. 7, 1870	F.	35	Gilder's wife	Aphonia	Left Vocal Cord	(Papilloma)
95	Mar. 9, 1870	F.	43	Coachman's wife.	Aphonia 23 years...	Right Ventricular Band.	Fasciculated Sarcoma.
96	Mar. 14, 1870	F.	27	Machinist ...	Aphonia	Posterior Wall of Larynx.	Papilloma...
97	Mar. 21, 1870	F.	29	Butcher's wife.	Great pain and effort in speaking, and Dysphonia.	Posterior Wall of Larynx.	Fibroma ...
98	May 10, 1870	M.	2½	Lighterman's child.	Aphonia from birth; great Dyspnœa.	The entire cavity of Larynx.	Benign Epithelial Growth.
99	May, 1870	M.	21	Gentleman	Dysphonia ...	Both Vocal Cords ...	Fibro-cellular Growth and Myxoma.
100	May 23, 1870	M.	51	Traveller ...	Dysphonia ...	Left Vocal Cord ...	(Papilloma)

Treatment.	Result.	References and Remarks.
Evulsion with tube-forceps and common laryngeal forceps.	Cure ...	Case operated on in the presence of Dr. Thorowgood, Mr. H. A. Reeves, and others.
Evulsion with common antero-posterior forceps.	Cure ...	Case seen by Surgeon-Major Wyatt, Dr. Farquhar, and others.
Evulsion with antero-posterior forceps.	Cure ...	Operation in presence of Dr. Haden.
Evulsion with antero-posterior forceps.	Cure ...	Growth removed in one operation, in presence of Dr. Henry Hubbard and Mr. Keene.
Evulsion with antero-posterior forceps.	Cure ...	Growth removed in one operation, in presence of Dr. Stage and Mr. Keene.
Evulsion with common laryngeal forceps.	Cure	Case known to Dr. Carre, of Blackheath.
Evulsion with tube-forceps ...	Cure ...	Pieces removed in presence of Mr. Keene and Dr. Collum.
Evulsion with common laryngeal forceps.	Cure ...	Case seen by Dr. Bowditch, of Boston, U.S.A., and Dr. Webb, of Cincinnati.
Tracheotomy	Death ...	On post-mortem examination, the Larynx was found to be blocked up by Growths.
Evulsion with common and antero-posterior forceps.	Cure ...	Case known to Dr. Thomson, of Devonport.
Evulsion with common forceps	Cure ...	Growth entirely removed, on first attempt, in presence of Mr. Sankey and Mr. Pugin Thornton.

APPENDIX D.

———o———

*** From this Table all Cases of *malignant disease,* as well as all Cases
of "*false excrescences,*" have been carefully excluded. Notwithstanding
the great pains which have been taken to search every publication
accessible in this country, it is quite possible that a few Cases may
have been accidentally overlooked. The Author is also aware that a
large number of other similar Cases have been treated, though they
have not yet been placed on record.

No. of Case.	Date.	Sex.	Age.	Occupation.	Symptoms.	Situation.	Pathological Nature.
1	July 20, 1860	M.	33	Merchant ..	Dysphonia, Dyspnœa, and Cough.	Anterior portion of left Vocal Cord.	Benign Epithelial Growth.
2	1860	F.	15	...	Aphonia and Dyspnœa.	One large Growth at anterior Commissure. A smaller one below Anterior Commissure.	Soft and granular Growth.
3	1860	F.	20	...	Aphonia	Left Vocal Cord and right Ventricle.	Numerous Warty Growths.
4	1860	M.	56		Hoarseness, Cough on deglutition. Sensation of foreign body in Throat.	Right Ventricle ...	Greyish-white Polypoid Growth.
5	Aug. 1860	F.	29	Lady ...	Aphonia and Dyspnœa.	Right Vocal Cord ...	Not stated
6	Oct. 1860	M.	24	Lapidary ...	Aphonia, Dyspnœa, and Pain.	Both Vocal Cords and Ventricular Bands.	Not stated
7	Oct. 27, 1862	F.	22	...	Aphonia and Cough	Right Vocal Cord ...	Not stated
8	1860	M.	43	Merchant ...	Hoarseness, persistent Cough, and burning Sensation in Throat.	Anterior Commissure	Pedunculated Mucous Polypus.
9	1860	M.	40	Merchant ...	Sensation of pressure and burning in Throat.	Anterior wall of Larynx.	Pedunculated Polypus.
10	Nov. 2, 1860	F.	20	Lady ...	Varying Aphonia...	Two Growths on left Vocal Cord.	Not stated
11	1860	M.	43	Tailor ...	Aphonia	Whole length of both Vocal Cords.	Papilloma...
12	Dec. 1860	M.	32	Merchant ...	Hoarseness and Cough.	Posterior wall of Larynx.	...
13	1860	F.	3½	...	Loud barking Cough, after Croup.	Posterior wall of Larynx.	Not stated
14	1860	M.	7		Aphonia, Dyspnœa, Stridor, and Cough.	Right Ventricle ...	Mucous Polypus.
15	1861	M.	48	Printer ...	Not mentioned ...	Anterior Commissure and right Vocal Cord.	Papilloma...
16	1861	F.	...		Not stated ...	The entire lining membrane of the Larynx.	Cellular-Sarcoma and Papilloma.

Treatment.	Result.	Operator.	Reference and Remarks.
Excision with scissors and forceps.	Cure	Dr. Lewin	*Deutsche Klinik*, 1862, p. 202. Growth returned in nine months, but patient refused further treatment.
Evulsion of portion with forceps. Subsequent cauterization with nitrate of silver.	Cure ...	Idem ...	*Ibid.*, 1862, p. 202.
Cauterization and evulsion with forceps.	Improvement	Idem ...	*Ibid.*, p. 203.
Cauterization ...	Cure	Idem ...	*Ibid.*, 1862, p. 243. This Growth was believed to originate from irritation of a fish-bone in the Larynx.
Destruction with caustic solutions.	Cure	Sir Duncan Gibb.	*Diseases of the Throat, &c.*, second edition, p. 143.
Destruction with caustic solutions.	Cure	Idem ...	*Ibid.*, p. 142.
Destruction with caustic solutions.	Cure	Idem ...	*Ibid.*, p. 144.
Destruction with caustic solutions.	Cure	Lewin ...	*Deutsche Klinik*, 1862, p. 244.
Cauterization for a fortnight. Sudden disappearance of tumour.	Cure	Idem ...	*Ibid.*, p. 245.
Destruction with caustic solutions.	Cure	Gibb ...	*Diseases of the Throat, &c.*, second edition, p. 145.
Partial evulsion with forceps.	Negative ...	Lewin ...	*Deutsche Klinik*, 1862, p. 258.
Cauterization ...	Negative ...	Idem ...	*Ibid.*, p. 259.
One application, under chloroform, of solution of nitrate of silver.	Cure	Idem ...	*Ibid.*, p. 245. It is scarcely possible that this could have been a case of true Growth.
Two successive cauterizations with solid nitrate of silver.	Cure	Idem ...	*Ibid.*, p. 257.
Evulsion with forceps ...	Temporary benefit.	Dr. Fauvel	*Du Laryngoscope au point de vue pratique.* Paris, 1861.
Tracheotomy. Division of Thyroid Cartilage, excision of Growth and of right Vocal Cord with knife.	Improvement	Dr. Rauchfuss.	*St. Petersburg mediz. Zeitsch.*, 1862, vol. vi. p. 44. *Death* two years later from Gangrene of Lungs. On *Autopsy*, communication between the middle of trachea and œsophagus. Right bronchus contained a chicken-bone.

No. of Case.	Date.	Sex.	Age.	Occupation.	Symptoms.	Situation.	Pathological Nature.
17	1861	...	...	...	Not stated ...	Not stated	Not stated
18	1861	M.	25	... ...	Not stated ...	Anterior Wall of Larynx.	
19	July 20, 1861	M.	48	Librarian ...	Aphonia and Cough	Below the anterior Commissure of Vocal Cords.	Fibroma ...
20	1861	M.	26	Gardener ...	Hoarseness, Cough, and Dyspnœa.	Right Ventricular Band.	Benign (?) Epithelial Growth.
21	1861	M.	14	Labourer ...	Aphonia and Dyspnœa.	Above Vocal Cords	Not stated
22	Jan. 14, 1862	M.	62	Merchant ...	Aphonia, occurring suddenly.	Both Vocal Cords, both Ventricular Bands, and Epiglottis.	Benign (?) Epithelioma.
23	1862	M.	29	Schoolmaster	Dysphonia ...	Right Vocal Cord and posterior Wall of Larynx.	Papilloma...
24	1862	M.	60	Apothecary	Hoarseness ...	Below anterior Commissure, and on left Vocal Cord.	Polypus ...
25	1862	M.	45	Gentleman	Hoarseness ...	Right Vocal Cord ...	Papilloma...
26	April 19, 1862	M.	31	Fireman ...	Aphonia ...	Posterior Wall of Larynx.	Not stated
27	Oct. 17, 1862	M.	27			Left Vocal Cord ...	Mucous Polypus.
28	1862	F.	20		Aphonia	Anterior Commissure of Vocal Cords and Ventricular Bands.	Not stated
29	1862	F.	39	Lady ...	Aphonia, Cough, and Oppression.	Two Growths below Vocal Cords.	Polypus ...
30	Nov. 1862	M.	37	Gentleman	Aphonia	Two Growths on left Vocal Cord.	Fibro-Epithelial Growth.
31	Dec. 10, 1862	M.	42	Gentleman	Hoarseness ...	Anterior Commissure	Ditto ...
32	Dec. 23, 1862	M.	47	Captain in Merchant Service.	Hoarseness ...	Between Vocal Cords and on left Vocal Cord.	Not stated
33	Dec. 1862	M.	37	Clergyman	Hoarseness and temporary Aphonia.	Left Vocal Cord ...	Fibroma ...

Treatment.	Result.	Operator.	Reference and Remarks.
Destruction by insufflation of powdered nitrate of silver.	Cure	Dr. Knauf	*Schmidt's Jahrbücher*, 1862, No. 6, p. 114. The tumour shrank, and was subsequently expelled.
Tracheotomy, March 26, 1861; Thyrotomy, Aug. 13, 1862.	Improvement	Dr. Gurdon Buck.	*New York Medical Journal*, May, 1865. Tracheotomy was performed for supposed abscess.
Excision with scissors ...	Cure	Prof. Bruns	*Die Laryngoskopie*. Tübingen, 1865, p. 57.
Evulsion with curved forceps.	Improvement; recurrence.	Messrs. Bracey and Bolton.	Dr. Russell, *On Laryngeal Diseases, &c.*, p. 16. 1864. Reprinted from *Brit. Med. Journal*.
Evulsion with écraseur ...	Great improvement.	Dr. Walker	*Lancet*, Nov. 9, 1861.
Evulsion with forceps, and subsequent cauterization.	Improvement	Prof. Czermak.	*Die Kehlkopfspiegel*, 1863, p. 117.
Excision with knife; application of nitrate of silver and galvano-cautery.	Improvement	Dr. Voltolini	*Die Anwendung der Galvanocaustic, &c.* Wien, 1867, p. 51. Case 4.
One excised with knife; the other destroyed by nitrate of silver and galvano-cautery.	Cure	Idem ...	*Ibid.*, p. 58. Case 6.
Excision with galvanic-cautery wire and laryngeal forceps.	Great improvement.	Idem	*Ibid.*, p. 56. Case 7.
Destruction with nitrate of silver and chromic acid.	Improvement	Dr. Louis Elsberg.	*Treatment of Morbid Growths within the Larynx*. Philadelphia, 1866. Case 2.
Evulsion with curved forceps, and subsequent cauterization.	Great improvement.	Dr. Weiss ..	*St. Petersburg medic. Zeits.*, 1864, p. 10. Growth recurred in 1863 and in March, 1864; was again completely removed by same means as before.
Destruction with galvanic-cautery.	Improvement	Voltolini ...	*Die Anwendung der Galvanocaustic*, p. 52. Case 5.
Excision with guillotine, in two sittings.	Cure	Dr. Charles Ozanam.	*Compte rendu de l'Académie des Sciences*, June 16, 1863.
Evulsion with écraseur ...	Cure	Gibb ...	*Diseases of the Throat, &c.*, second edition, p. 147.
Evulsion with écraseur, under chloroform.	Cure ...	Idem ...	*Ibid.*, p. 149.
Evulsion with écraseur ...	Cure	Idem ...	*Ibid.*, p. 150.
Excision with double-edged knife.	Cure	Bruns	*Die Laryngoskopie, &c.* Tübingen, 1865, p. 270.

No. of Case.	Date.	Sex.	Age.	Occupation.	Symptoms.	Situation.	Pathological Nature.
34	Nov. 28, 1862	M.	45	...		Anterior half of both Vocal Cords.	Cauliflower Polypi.
35	1863	M.	31	Priest	Hoarseness ...	Anterior Commissure	Not stated
36	Jan. 1863	M.	22	Railway official.	Hoarseness and slight Hæmoptysis.	Right Vocal Cord ...	Fibro-cellular Polypus.
37	Feb. 1863	M.	15		Aphonia and Attacks of Suffocation.	Whole inner surface of Larynx.	Papilloma...
38	Feb. 1863	M.	51		Hoarseness ...	Posterior Wall of Larynx.	Not stated
39	Mar. 1863	M.	40		Aphonia	Epiglottis, right Vocal Cord, and left Arytenoid Cartilage.	Papilloma...
40	1863	F.				Epiglottis	Fibroma (as large as a walnut).
41	Mar. 26, 1863	F.	39			Internal surface of left Ary-epiglottic Fold.	Fibroma Polypus as large as a filbert.
42	Mar. 29, 1863	F.	...			Vocal Cord	Nodulated Tumour.
43	Mar. 1863	M.	...		Hoarseness and Attacks of Suffocation.	Several Growths filling the Larynx.	Polypi ...
44	1863	M.	...			Vocal Cord	Firm lobulated Polypus.
45	Mar. 29, 1863	F.	25	Lady ...	Aphonia	Posterior Wall of Larynx.	Not stated
46	Mar. 1863	F.	16		Aphonia and Dyspnœa.	Posterior Wall of Larynx.	Not stated
47	1863	M.	Boy		Aphonia	Left Vocal Cord ...	Two Polypi
48	May, 1863	M.	28	Warehouseman.	Almost complete Aphonia.	Left Vocal Cord and left Ventricular Band.	Fibroma ..
49	1863	F.	...	Lady ...	Speaking voice changed; singing voice lost.	Right Vocal Cord ...	Polypus ...
50	1863	M.	Mid. Age	Coal-merchant.	Dysphonia and "Catch" in breathing.	Left Vocal Cord	Five small Polypi.

Treatment.	Result.	Operator.	Reference and Remarks.
Evulsion with curved forceps; subsequent cauterization.	Improvement	Fauvel ...	*Gazette hebdomadaire*, May 29, 1863.
Destruction with caustic solutions.	Improvement	Dr. Türck...	*Klinik der Krankheiten des Kehlkopfes*, p. 311.
Evulsion with crushing forceps and destruction with caustic solutions.	Improvement	Dr. Stoerk	Wagner's *Archiv der Heilkunde*, p. 238.
Tracheotomy. Subsequent evulsion through the mouth and cauterization.	Cure	Rauchfuss...	Dr. Causit's *Etudes sur les Polypes du Larynx*. Paris, 1867, Case 39, p. 148. Patient was a deaf mute.
Evulsion with écraseur ...	Improvement	Gibb ...	*Diseases of the Throat, &c.*, second edition, p. 151.
Cauterization; subsequent evulsion and crushing.	Negative ...	Dr. Lindwurm.	*Medical Times and Gazette*, April 5, 1862 (Munich Correspondent).
Partial removal with écraseur; subsequent excision with knife.	Cure	Professor Langenbeck.	*Gazette hebdomadaire*, March 13, 1863.
Evulsion with wire-écraseur.	Cure	Dr. Trélat...	*Ibid.*, May 1, 1863. Growth seen without laryngoscope.
Evulsion with forceps and subsequent cauterization.	Cure	Rauchfuss...	*St. Petersburg medizin. Zeitschrift*, 1864, p. 144.
Tracheotomy. Growths subsequently removed with forceps through the mouth.	Improvement	Idem ...	*Ibid.*, p. 153. Patient continued to wear the tube. Breathing relieved, but hoarseness continued.
Unsuccessful attempts being followed by great Dyspnœa, tracheotomy performed. Subsequent evulsion of Growth through mouth.	Improvement	Idem ...	*Ibid.*, p. 45. Patient continued to wear the tube. Breathing relieved, but hoarseness continued.
Evulsion with écraseur ...	Great improvement.	Gibb ...	*Diseases of the Throat, &c.*, second edition, p. 152.
Evulsion with écraseur ...	Cure	Idem ...	*Ibid.*, p. 153.
Evulsion with loop-écraseur.	Cure	Elsberg ...	*Treatment of Morbid Growths within the Larynx*. Philadelphia, 1866. Case 3.
Partial destruction with caustic solutions.	Great improvement.	Idem ...	*Ibid.*, Case 5. "A sufficiently loud and clear voice restored." Traces of the Growth remained.
Evulsion with forceps ...	Cure	Idem ...	*Ibid.*, Case 4.
Evulsion of the largest Growth; cauterization of remainder.	Cure	Idem ...	*Ibid.*, Case 6.

No. of Case.	Date.	Sex.	Age.	Occupation.	Symptoms.	Situation.	Pathological Nature.
51	1863	M.	22			Larynx and Epiglottis	Benign (?) Epithelial Growth.
52	1863	M.	30	Tinsmith ...	Dysphonia and Dyspnœa.	Posterior Wall of Larynx.	Not stated
53	1863	M.	40		Hoarseness from childhood ; Cough for 3 weeks.	Cushion of Epiglottis	Pedunculated Tumour.
54	1863	F.	35		Dysphonia and fatigue of Voice.	Both Vocal Cords ...	Not stated
55	1863	M.	11		Dysphagia, Dysphonia, and Dyspnœa.	Epiglottis ...	Cystic Growth.
56	1863	M.	21		Dyspnœa	Upper Orifice of Larynx.	Fibro-cellular Polypi.
57	May 2, 1863	F.	38		Aphonia and Dyspnœa.	Right Vocal Cord ...	Cystic Growth.
58	May 6, 1863	M.	33	Merchant ...	Hoarseness ...	Right Vocal Cord ...	Fibroma ...
59	1863	F.	33	...	Aphonia	Right Vocal Cord ...	Papilloma...
60	June 6, 1863	M.	34	Priest ...	Hoarseness, Pain, and Sensation of foreign body.	Right Vocal Cord ..	Fibroma ...
61	June 24, 1863	M.	23	Merchant ...	Hoarseness ...	Right Vocal Cord ...	Fibroma ...
62	July 1, 1863	M.	31		Hoarseness, Pain in speaking.	Right Vocal Cord ...	Fibroma ...
63	July 17, 1863	F.	26	...	Hoarseness, Dyspnœa.	Left Ventricular Band.	Fibroma ...
64	July, 1863	M.	43	...		Posterior Wall of Larynx, filling half the Glottis.	Fibro-cellular Polypus.
65	Aug. 11, 1863	F.	24	...	Dysphonia and Dyspnœa.	Superior Orifice of Larynx.	Papilloma...
66	Aug. 1863	M.	...		Dyspnœa and Stridor.	One of the Vocal Cords.	Not stated
67	Oct. 25, 1863	M.	28		...	Left Vocal Cord ...	Warty excrescence, composed of fibrous tissue and epithelium.

Treatment.	Result.	Operator.	Reference and Remarks.
Partial evulsion with curved forceps.	Improvement	Czermak ...	*Medical Times and Gazette*, May 30, 1863.
Destruction with nitric acid and other applications.	Improvement	Elsberg ...	*Treatment of Morbid Growths within the Larynx.* Philadelphia, 1866. Case 7. Dyspnœa relieved; result as to voice not stated.
Evulsion with wire loop...	Cure	Idem ...	*Ibid.*, Case 8.
Abscission with Tobold's scissors.	Cure	Idem ...	*Ibid.*, Case 9.
Incision with bistoury ...	Cure	Mr. Durham	*Medico-Chirurgical Transactions*, 1863.
Sub-hyoid Laryngotomy and removal of Growth.	Cure	Dr. Follin...	*Archives générales de Médecine.* Feb. 1867.
Evulsion with écraseur ...	Cure	Gibb ...	*Diseases of the Throat, &c.*, second edition, p. 154.
Incision with knife, evulsion with wire loop.	Cure	Bruns ...	*Die Laryngoskopie, &c.* Tübingen, 1865, p. 278.
Destruction with crushing-forceps, followed by cauterization.	Cure	Türck ...	*Klinik der Krankheiten des Kehlkopfes*, p. 304.
Incision with knife, evulsion with forceps.	Cure	Bruns ...	*Die Laryngoskopie, &c.* Tübingen, 1865, p. 287.
Incision with knife, evulsion with wire loop.	Cure	Idem ...	*Ibid.*, p. 298.
Incision with knife, evulsion with forceps.	Cure	Idem ...	*Ibid.*, p. 301.
Preliminary Tracheotomy. Excision with scissors, evulsion with forceps and wire loop.	Cure	Idem ...	*Ibid.*, p. 307.
Tracheotomy and division of Thyroid and Cricoid Cartilages. Growth evulsed, and application of galvanic-cautery.	No benefit ...	Dr. Busch	*Beobachtungen zur innern Klinik*, von Carl Binz. Bonn, 1864, p. 108. Recurrence of Growth.
Tracheotomy, division of Thyroid and Cricoid Cartilages, and of Epiglottis; and removal of Growths.	Death ...	Dr. Boeckel	*Extrait de la Thèse de Swebel*, Strasbourg, 1866, quoted by Planchon in his *Faites cliniques de Trachéotomie*, Paris, 1869. Patient died two months later of Pneumonia.
Evulsion with écraseur ...	Cure	Gibb ...	*Diseases of the Throat, &c.*, second edition, p. 155. No details given.
Evulsion with guillotine...	Cure	Dr. Semeleder.	*Wiener Medizinalhalle.* 1864.

No. of Case.	Date.	Sex.	Age.	Occupation.	Symptoms.	Situation.	Pathological Nature.
68	Nov. 1, 1863	F.	...		...	Left Vocal Cord and right Arytenoid Cartilage.	Three separate Polypi.
69	Nov. 17, 1863	M.	52	... ...	Aphonia, Dyspnœa, Dysphagia.	Ary-epiglottic Fold...	Large Fibroma.
70	1863	M.	22	Railway Porter.	Dysphonia ...	Right Vocal Cord ...	Papilloma...
71	1863	F.	26	... ...	Aphonia, attacks of Suffocation.	Right Vocal Cord ...	Pedunculated Polyp.
72	1863	F.	44	Charwoman	Hoarseness, Suffocating Cough, Sensation of foreign body in Larynx.	Ary-epiglottic Fold...	Polypus ...
73	Nov. 1863	M.	29	Vocalist ...	Dysphonia ...	Right Vocal Cord ...	Fibroma ...
74	Jan. 1, 1864	F.	47		Dysphonia ...	Left Ventricular Band	Polypus ...
75	Jan. 2, 1864	F.	29	...	Aphonia	Ventricular Band ...	Polypus ...
76	Jan. 25, 1864	F.	21	Lady ...	Aphonia and severe hacking Cough.	Entire cavity of Larynx, except posterior portion of right Vocal Cord.	Fibro-Epithelial Growth.
77	Mar. 1864	M.	10		Hoarseness, Dyspnœa.	Left Ventricle ...	Papilloma...
78	Mar. 1864	M.	49	Cabinet-maker.	Voice extremely husky and feeble.	Right Vocal Cord ...	Benign Epithelial Growth.
79	1864	F.	55	Servant ...	Hoarseness, Sensation as of a hair in Larynx.	Left Ventricle ...	Not stated
80	Mar. 1864	F.	...	... ...	Hoarseness and occasional Aphonia.	Left Vocal Cord	Fibro-epithelial Growth.
81	April, 1864	M.	...	Railway official.	Aphonia	Below left Vocal Cord	Papilloma...
82	1864	M.	...	Singer ...	Hoarseness, Cough, Sensation of foreign body in Larynx.	Right Vocal Cord ...	Polypus ...
83	April, 1864	F.	48	...	Dysphonia, subsequent Aphonia and Pain.	Left Vocal Cord	Papilloma...
84	May, 1864	M.	64	Physician ...	Dysphonia ...	Left Ventricle ...	Cystic Growth.
85	May, 1864	M.	49	Lithographer	Dysphonia ...	Left Vocal Cord ...	Papilloma...

Treatment.	Result.	Operator.	Reference and Remarks.
Evulsion with various instruments.	Cure	Dr. Seme-leder.	*Wiener Medizinalhalle,* 1864.
Transverse incision through Thyro-hyoid membrane, and Thyrotomy. Tracheotomy subsequently performed.	Death ...	M. Debrou	*Gazette des Hôpitaux,* 1864, No. 46. Patient died 7 days after operation, with metastatic abscesses in both lungs.
Destruction with crushing forceps.	Great improvement.	Dr. Theodor Stark.	*Archiv der Heilkunde,* p. 237 et seq. "But slight hoarseness remained."
Excision with guarded knife.	Cure	Stoerk ...	*Wiener medizin. Wochenblatt,* 1863, No. 44.
Evulsion with wire loop and cauterization.	Cure	Trélat ...	*Gazette des Hôpitaux,* May 2, 1863.
Excision with knife, subsequent cauterization.	Cure	Bruns ...	*Die Laryngoskopie, &c.* Tübingen, 1865, p. 354.
Destruction with galvanic-cautery, after various unsuccessful treatment.	Improvement	Idem ...	*Ibid.,* p. 365.
Excision with knife ..	Improvement	Stoerk ...	*Wiener Wochenbl.,* xx. p. 22, 1864.
Evulsion with cutting forceps.	Negative ...	Elsberg ...	*Treatment of Morbid Growths within the Larynx.* Philadelphia, 1866. Tumour recurs as fast as it is removed; division of Thyroid Cartilage contemplated.
Evulsion with wire loop...	Improvement	Bruns ...	*Die Laryngoskopie, &c.* Tübingen, 1865, p. 370.
Evulsion with écraseur ...	Improvement	Dr. George Johnson.	*Transactions of Medico-Chirurgical Society,* vol. li. Case 1. "Voice still feeble and husky."
Destruction by one application of solid nitrate of silver.	Cure	Türck ...	*Klinik der Krankheiten des Kehlkopfes,* p. 310. It is very improbable that this was a case of true Growth.
Excision with cutting forceps.	Cure	Elsberg ...	*Treatment of Morbid Growths within the Larynx.* Philadelphia, 1866.
Excision with curved scissors and cauterization.	Improvement	Dr. Otto Prinz.	*Archiv für Heilkunde,* 1866, p. 220.
Scarification, cauterization, and application of solutions of alum.	Cure	Idem ...	*Ibid.,* p. 222.
Evulsion and galvanic cautery.	Cure	Bruns ...	*Die Laryngoskopie, &c.* Tübingen, 1865, p. 389.
Puncture with knife ...	Cure	Idem ...	*Ibid.,* p. 383.
Evulsion with wire loop...	Improvement	Idem ...	*Ibid.,* p. 400.

No. of Case.	Date.	Sex.	Age.	Occupation.	Symptoms.	Situation.	Pathological Nature.
86	Summer of 1864	M.	26	Dispenser ...	Hoarseness ...	Right Vocal Cord ..	Benign Epithelial Growth.
87	June 6, 1864	M.	39	Shepherd ..	Hoarseness ..	Right Vocal Cord ...	Fibroma ...
88	July 20, 1864	M.	59	Civil Service official.	Dysphonia ...	Right Vocal Cord ..	Fibroma ...
89	Oct. 10, 1864	M.	5		Aphonia	Entire surface of Larynx and upper part of Trachea.	Papilloma...
90	Oct. 1864	M.	47	Gentleman	Dysphonia ...	Both Vocal Cords ...	Papilloma...
91	Oct. 12, 1864	M.	59	Gentleman	Dysphonia ...	Right Vocal Cord ...	Fibroma ...
92	Oct. 31, 1864	F.	16	...	Hoarseness, 12 yrs., Aphonia, and Dyspnœa.	Both Ventricles ...	Polypoid mass.
93	Dec. 1, 1864	M.	9½	...	Hoarseness and attacks of Suffocation.	Region of Vocal Cords	Polypus ...
94	1864	M.	...	Consul ...	Aphonia, with great irritability of Palate.	Left Ventricular Band	Polypus ...
95	Dec. 8, 1864	F.	6		Dyspnœa, and attacks of Suffocation since three months old.	Entire cavity of Larynx and part of Trachea.	Cauliflower Excrescence
96	1865	M.	34	Professor ...	Hoarseness ..	Lower surface of Right Vocal Cord.	Pedunculated Sarcoma.
97	1865	M.	...	Soldier ...	Aphonia	Right Ventricle ...	Fibroma ...
98	May 19, 1865	M.	59	Manager of Estates.	Hoarseness ...	Right Vocal Cord ...	Benign Epithelial Growth.
99	June 29, 1865	M.	50	Lay-Agent	Hoarseness ...	Anterior insertion of Vocal Cords.	Vascular Cyst.
100	1865	M.	51	Clergyman	Hoarseness and Cough.	Right Vocal Cord ...	Two small Polypi.
101	June 29, 1865	M.	52		Aphonia	Left Vocal Cord ...	Sarcoma, rich in connective tissue.

Treatment.	Result.	Operator.	Reference and Remarks.
Evulsion with écraseur ...	Cure	Johnson ...	*Transactions of Medico-Chirurgical Society*, vol. li. Case 4.
Destruction with galvanic-cautery loop.	Cure ...	Voltolini ...	*Die Anwendung der Galvano-caustic, &c.* Wien, 1867, p. 57.
Excision with knife, and cauterization.	Cure	Bruns ...	*Die Laryngoskopie, &c.* Tübingen, 1865, p. 465.
Removal with wire loop and scraper, and various caustics; preceded by Tracheotomy.	Improvement	Idem ...	*Ibid.*, p. 322. Tracheotomy performed before patient came under Dr. Bruns' care.
Excision with knife and scissors.	Cure	Idem ...	*Ibid.*, p. 410.
Incision with knife; evulsion with wire loop.	Cure	Idem ...	*Ibid.*, p. 415.
Tracheotomy, Thyrotomy, and removal of Growth.	Cure	Drs. Lewin and Ulrich	*Deutsche Klinik*, 1865, No. 52, p. 510. Canula removed on third day.
Tracheotomy. Subsequent excision with wire écraseur.	Improvement	M. Giraldes	*Gazette des Hôpitaux*, No. 140, 1867. Patient wore canula for four years. Diagnosis made with finger without aid of laryngoscope.
Excision with curved scissors, forceps, and cauterization.	Improvement	Dr. Tobold	*Berlin. klin. Wochenschrift*, 1864, No. 40, p. 386.
Dec. 4th, Tracheotomy. Feb. 26, 1865, Division of Thyroid and Cricoid Cartilages and removal of Growth with scissors. In November, operation repeated, and cauterization with chromic acid.	Improvement	Prof. Gouley	*New York Medical Journal*, September, 1867, p. 473. In July, 1867, patient spoke in a loud very distinct whisper, breathing quite natural.
Evulsion with forceps ...	Improvement	Türck ...	*Klinik der Krankheiten des Larynx*, p. 576.
Excision with Bruns' knife and cauterization.	Cure	Prinz ...	*Archiv für Heilkunde*, p. 223.
Evulsion with écraseur ...	Great improvement.	Johnson ...	*Transactions of Medico-Chirurgical Society*, vol. li. Case 3. No recurrence in July, 1867, but cord uneven, and voice still gruff.
Excision of Cyst with écraseur.	Cure	Idem ...	*Ibid.*, Case 2.
Destruction with solid nitrate of silver.	Improvement	Voltolini ...	*Deutsche Klinik*, 1865, p. 65.
Excision with knife, and crushing forceps; subsequent cauterization.	Cure	Dr. Schroetter.	*Wochenblatt der Wiener Aerzte*, 9th August, 1865, No. 34.

No. of Case.	Date.	Sex.	Age.	Occupation.	Symptoms.	Situation.	Pathological Nature.
102	1865	M.	43	...	Hoarseness ...	Right Vocal Cord ...	Mucous Polypus.
103	1865	F.	54	...	Hoarseness ...	Left Vocal Cord ...	Fibroma ...
104	July, 1865	F.	42	...	Aphonia, Dyspnœa, Cough.	Numerous Growths in Larynx.	Papilloma...
105	July 24, 1865	F.	44		Aphonia, Dyspnœa, Stridor, and Dysphagia.	Ventricle, and filling the Larynx.	Benign (?) Epithelioma; five tumours in all.
106	1865	M.	48	Tradesman	Hoarseness and Cough.	Right Vocal Cord ...	Tumour of inflammatory infiltration.
107	1865	M.	57	Locksmith	Attacks of Suffocation.	Both Vocal Cords ..	Several Polypi.
108	1865	M.	42	Officer ...		Right Ventricular Band and left Vocal Cord.	Benign (?) Epithelioma.
109	1865	M.	33	Actor ...		Right Vocal Cord ...	Pedunculated Polypus.
110	Dec. 21, 1865	M.	38	Police Inspector.	Hoarseness ...	Left Vocal Cord and Ventricle.	Papilloma...
111	Dec. 1865	M.	39	Innkeeper...	Aphonia	Anterior Commissure and both Vocal Cords.	Papilloma...
112	1865	F.	16		Hoarsenes and Stridulous breathing.	Anterior Commissure of Vocal Cords.	Polypus ...
113	1865	F.	54		Hoarseness ...	Left Vocal Cord ...	Polypus ...
114	1865	M.	...	...	Aphonia and severe Dyspnœa.		Polypoid Growth.
115	1865	M.	48	Inspector of prisons.	Hoarseness, Dyspnœa.	Anterior part of right Vocal Cord.	Polypus ...
116	Nov. 3, 1865	M.	40	Schoolmaster.	Hoarseness, Dyspnœa, Sensation of foreign body in Larynx.	Left Vocal Cord ...	Fibroma ...
117	Dec. 3, 1865	M.	35	Physician ...	Dysphonia ...	Both Vocal Cords and Ventricular Bands.	Papilloma...
118	Dec. 2, 1865	M.	38	Police inspector.	Hoarseness ...	Left Vocal Cord and left Ventricle.	Papilloma...
119	Jan. 4, 1866	M.	52	Mason ...	Hoarseness and tickling sensation.	Left Vocal Cord ...	Papilloma....

tment.	Result.	Operator.	Reference and Remarks.
ith écraseur g forceps.	Cure	Türck ...	*Klinik der Krankheiten des Kehl-kopfes*, p. 299.
ith crushing	Cure	Idem ...	*Ibid.*, p. 299.
double-edged fe.	Cure	Dr. Gottstein	*Berlin. klin. Wochenschrift*, No. 46, 1865.
', division of and Cricoid and excision is with flat	Cure	Prof. von Balassa.	*Wiener medizin. Wochenschrift*, Nov. 1868. Canula was removed, and voice continued clear.
with crushing	Improvement	Türck ...	*Klinik der Krankheiten des Kehl-kopfes*, p. 478.
Excision with nd cauteriza·	Improvement	M. Koeberlé	*Gazette des Hôpitaux*, June 13, 1865.
h knife, and n.	Improvement	Dr. Emrich Navratil.	*Berlin. klin. Wochenschrift*, 1868, No. 48.
h knife, and n.	Cure	Idem ...	*Ibid.*, Vocal cord was wounded in operation. Subsequent cure by faradization.
vith galvanic· e loop.	Improvement	Voltolini ...	*Archiv für Heilkunde*, 1866, p. 63.
th crushing	Cure	Türck ...	*Klinik der Kehlkopfkrankheiten*, p. 578. Growth recurred in 1866, and was treated by excision and cauteriza-tion.
nd excision...	Improvement	Gilewski ...	*Wiener medizin. Wochensch.*, June 28, 1865, p. 142.
h crushing	Cure	Türck ...	*Klinik der Kehlkopfkrankheiten*, p. 309.
	Improvement	Stoerk ...	*Wiener Jahrbücher der k. k. Gesell-schaft der Aerzte*, No. 22, 1865.
and removal ular orifice.	Cure	Dr. Burow, senior.	*Deutsche Klinik*, vol. xvii. p. 165.
knife and	Cure	Bruns ...	*Polypen des Kehlkopfes*. Tübingen, 1868, p. 8.
ith wire loop, d cauteriza·	Improvement	Idem ...	*Ibid.*, p. 2.
ith galvanic· loop.	Negative ...	Voltolini ...	*Die Anwendung der Galvano-caustic, &c.* Wien, 1867, p. 63. Patient still under treatment at time of publication.
ith galvanic· loop.	Cure	Idem ...	*Ibid.*, p. 59.

No. of Case.	Date.	Sex.	Age.	Occupation	Symptoms.	Situation.	Pathological Nature.
120	1866	M.	55	Watchmaker	Hoarseness ...	Right Vocal Cord ...	Pedunculated Sarcoma.
121	April, 1866	M.	56	Tax-agent ...	Aphonia	Right Vocal Cord ...	Myxoma ...
122	April 28, 1866	M.	48	Teacher ...	Hoarseness and Pain.	Right Vocal Cord ...	Fibroma ...
123	Aug. 1866	M.	74	Clockmaker	Hoarseness, Dyspnœa, and Stridor.	Left Vocal Cord ...	Adenoma ...
124	April, 1866	F.	32	Charwoman	Aphonia, Dyspnœa, Cough, and Pain.	Right Vocal Cord ...	Soft, nodular Tumour, with short Peduncle.
125	1866	M.	38	... ...	Dysphonia ...	Right Vocal Cord ...	Two small, whitish, fibrous Polypi.
126	1866	M.	32	Chimney-sweep.	Hoarseness, Cough, and Dyspnœa.	Right Ventricle ...	Benign (?) Epithelioma.
127	1866	M.	47	...	Aphonia and Dyspnœa.	Right Vocal Cord filling the Glottis.	Papilloma...
128	1866	M.	46	Mechanic ...	Aphonia	Right Vocal Cord ...	Fibro-plastic growth.
129	1866	F.	15	...	Aphonia	Left Vocal Cord ...	Cystic Tumour.
130	1866	M.	35	...	Severe Dyspnœa	Right Vocal Cord ...	Polypus ...
131	1866	...	...	...	Aphonia and Dyspnœa.	Right Vocal Cord ...	Papilloma...
132	1866	M.	38	Teacher ...	Aphonia	Right Vocal Cord ...	Polypus ...
133	1866	M.	36	Dyer ...	Hoarseness, Dyspnœa, and Cough.	Beneath Anterior Commissure.	Polypus ...
134	1866	M.	54	Printer ...	Aphonia and Dyspnœa.	Whole lining membrane of Larynx.	Papilloma...
135	1866	F.	28	Single Lady	Hoarseness, Pain, and Dysphagia.	Left side of Larynx...	Papilloma...
136	Dec. 7, 1866	F.	19	Peasant Girl	Aphonia	Anterior Commissure	Papilloma...
137	1866	M.	42	Merchant ...	...	Left Vocal Cord ...	Soft Polypus
138	1866	M.	39	Merchant ...	Aphonia and Dyspnœa.	Right Vocal Cord ...	Fibro-epithelial growth.

Treatment.	Result.	Operator.	Reference and Remarks
Excision with knife ...	Improvement	Türck ...	*Klinik der Kehlkopfkrankheiten*, p. 577.
Evulsion with wire loop...	Cure	Bruns ..	*Polypen des Kehlkopfes.* Tübingen, 1868, p. 17.
Partially removed with knife.	Improvement	Idem ...	*Ibid.*, p. 23.
Incision ; evulsion with forceps and wire loop.	Cure ...	Idem ...	*Ibid.*, p. 30.
Evulsion with forceps, and cauterization.	Cure ...	Dr. Ludwig Mayer.	*Wiener medizin. Wochenschrift*, 1866, No. 36, p. 375.
Cauterization with nitrate of silver and acetic acid.	Improvement	Dr. Krishaber.	Dr. Causit's *Etudes sur les Polypes du Larynx*, 1867, p. 150.
Thyrotomy (without preliminary Tracheotomy), and excision of Growth.	Cure	Balassa ...	*Wiener medizin. Wochenschrift*, 1868, No. 93.
Excision with doubleedged polyp-knife.	Cure	Dr. Schnitzler.	*Wiener medizin. Presse*, 1866, No. 50.
Excision with doubleedged knife, and caustics.	Improvement	Dr. Henry Oliver.	*American Journal of Medical Science*, 1867, p. 115. A recurrence took place.
Excision with knife ...	Cure	Dr. Merkel	*Deutsche Klinik*, No. 29, 1866.
Destruction with galvaniccautery loop.	Cure	Schnitzler ..	*Wiener Wochenblatt der Aerzte*, 1867, No. 3.
Excision with doubleedged knife.	Cure ...	Idem ...	*Wiener medizin. Presse*, 1867, Nos. 20, 26.
Evulsion with wire loop ...	Cure ...	Professor Gerhardt.	*Jahresberichte über die Fortschritte.* Virchow, vol. ii. p. 127.
Excision with wire loop, knife, and hooks.	Improvement	Idem ...	*Ibid.*
Destruction with various instruments, galvanocautery, &c.	Negative ...	Voltolini ...	*Die Anwendung der Galvanocaustic*, &c. Wien, 1867, p. 50.
Excision with knife, cauterization and galvanic cautery.	Negative ...	Idem ...	*Ibid.*, p. 51.
Evulsion with wire loop ...	Cure	Bruns ...	*Polypen des Kehlkopfes.* Tübingen, 1868, p. 40.
Excision with knife, and subsequent cauterization.	Improvement	Navratil ...	*Berlin. klin. Wochenschrift*, 1868, Nov. 30, p. 489.
Excision with Tobold's scissors and écraseur.	Cure	Dr. A. Ruppaner ...	*New York Medical Journal*, vol. x. No. 4, p. 337.

No. of Case.	Date.	Sex.	Age.	Occupation.	Symptoms.	Situation.	Pathological Nature.
139	1866	M.	...	Ragman ...	Hoarseness and Dyspnœa.	Right Vocal Cord ...	... ..
140	Feb. 14, 1867	F.	21		Aphonia	Both Vocal Cords ...	Papilloma...
141	Feb. 1867	M.	68	Surgeon ...	Hoarseness and Cough.	Right Vocal Cord ...	Benign Epithelial Growth.
142	1867	M.		...	Aphonia	Not stated ...	Papilloma...
143	1867	M.	34	...	Severe Dyspnœa ...	Posterior Wall of Trachea.	Polypus ...
144	Mar. 1867	M.	37	Carpenter...	Aphonia	Both Vocal Cords and Ventricular Bands.	Papilloma...
145	Mar. 7, 1867	M.	38	Judge ...	Hoarseness ...	Left Vocal Cord ...	Papilloma...
146	April, 1867	F.	47		Aphonia, Dyspnœa, sudden attacks of Suffocation.	Under surface of Right Vocal Cord.	Polypus ...
147	April 30, 1867	F.	19	... ·	Aphonia, Dyspnœa, and attacks of Suffocation.	Below Vocal Cords...	Whitish granular Epithelioma of hard consistence.
148	April, 1867	M.	50	Merchant ...	Aphonia	Both Vocal Cords and Ventricular Bands.	Papilloma...
149	May 22, 1867	M.	22	Merchant ...	Aphonia	Both Vocal Cords and Ventricular Bands.	Papilloma...
150	May 22, 1867	M.	30	Chemist ...	Hoarseness ..	Left Vocal Cord ...	Fibroma ...
151	May 1867	M.	8	...	Hoarseness and slight Dyspnœa.	Right Vocal Cord ...	Benign Epithelial Growth.
152	May 20, 1867	M.	...		Aphonia and Dyspnœa.	Not stated	Fibroma ...
153	June, 1867	M.	...		Aphonia	Right Vocal Cord ...	Papilloma...
154	June 16, 1867	M.	36	Clergyman	Hoarseness ...	Both Vocal Cords ...	Polypus
155	June, 1867	M.	66	Farmer ...	Hoarseness ...	Right Vocal Cord ...	Papilloma...
156	Aug. 1867	F.	25	Single lady	Hoarseness ...	Left Arytenoid region	Lipoma, with fibroid base.
157	1867	M.	34	Teacher	Hoarseness and Pain.	Left Vocal Cord	Fibroma ...

Treatment.	Result.	Operator.	Reference and Remarks.
Tracheotomy, and subsequent destruction of growth with galvano-cautery.	Improvement	Dr. Venturini.	*Rivista clinica,* April, 1867; and *Il Galvano - caustico nelle Mal. della Laringe.* Bologna, 1867, 127.
Excision with polyp-knife and cauterization.	Cure ...	Tobold ...	*Berlin. klin. Wochenblatt,* 1869, Nos. 3 and 4.
Excision with écraseur ...	Cure ...	Johnson ...	*Transactions of Medico-Chirurgical Society,* vol. li. Case 5.
Excision with knife and cauterization.	Cure ...	Stoerk ...	*Wiener medizin. Wochenblatt,* March, 27, 1867.
Excision with forceps and application of astringents	Cure ...	Schroetter	*Wiener Zeitschrift der Aerzte,* 1867, p. 405.
Evulsion with wire loop ...	Improvement	Bruns ..	*Polypen des Kehlkopfes.* Tübingen, 1868, p. 49.
Evulsion with wire loop, scraping, and cauterization.	Cure ...	Idem ...	*Ibid.,* p. 44.
Evulsion with curved forceps and cauterization.	Not stated ...	Dr. J. S. Cohen.	*American Medical Journal,* April, 1867, p. 404.
Tracheotomy, division of Thyroid and Cricoid Cartilages, and excision of Growth.	Cure ...	Balassa ...	*Wiener medizin. Wochenschrift,* November 11, 1868.
Evulsion with forceps and wire loop.	Improvement	Bruns ...	*Polypen des Kehlkopfes.* Tübingen, 1868, p. 54.
Evulsion with wire loop and cauterization.	Improvement	Idem ...	*Ibid.,* p. 59.
Excision with knife ...	Cure ...	Idem ...	*Ibid.,* p. 64.
Excision with écraseur ...	Improvement	Johnson ...	*Transactions of Medico-Chirurgical Society,* vol. li. Case 6. "Voice somewhat reedy."
Tracheotomy and removal of Growth with forceps.	Cure ...	Dr. Fournié	*Gazette des Hôpitaux,* June 27, 1867.
Excision with knife and cauterization.	Improvement	Tobold ...	*Berlin. klin. Wochenschrift,* Dec. 30, 1867.
Excision with wire loop and knife.	Cure ...	Bruns ...	*Polypen des Kehlkopfes.* Tübingen, 1868, p. 67.
Evulsion with wire loop...	Cure ...	Idem ...	*Ibid.,* p. 77.
Destruction with galvanic-cautery, and excision with scissors and forceps.	Improvement	Idem ...	*Ibid.,* p. 84.
Excision with knife ...	Improvement	Bruns ...	*Polypen des Kehlkopfes.* Tübingen, 1868, p. 80.

No. of Case.	Date.	Sex.	Age.	Occupation.	Symptoms.	Situation.	Pathological Nature.
158	Sept. 1867	M.	17	Engraver ...	Attacks of Suffocation.	Below the Vocal Cords	Polypus ...
159	1867	F.	46		Obstinate and severe Cough.	Epiglottis ...	Melanotic Growth.
160	Sept. 1867	M.	16	... ...	Aphonia	Both Vocal Cords ...	Papilloma...
161	Oct. 1867	M.	48	Gentleman	Hoarseness ...	Right Vocal Cord ...	Benign Epithelial Growth.
162	Oct. 27, 1867	M.	32	Officer	Hoarseness ...	Left Vocal Cord ...	Fibroma ...
163	Oct. 1867	M.	26	Banker	Hoarseness ...	Right Vocal Cord ...	Fibroma ...
164	Nov. 11, 1867	F.	13		Aphonia and Dyspnœa.	Both Vocal Cords and Ventricular Bands.	Papilloma...
165	Nov. 1867	M.	32	Jeweller ...	Aphonia	Both Vocal Cords ...	Papilloma...
166	Nov. 1867	M.	48	Weaver	Aphonia ...	Anterior Commissure	Fibroma, with extensive ulceration of Larynx.
167	Nov. 1867	M.	63		Hoarseness, Cough, and Dyspnœa.	Entire lining membrane of Larynx.	Papilloma..
168	1867	M.	14		Hoarseness and Dyspnœa.	Right Vocal Cord ...	Papilloma...
169	Nov. 30, 1867	M.	59	Lawyer ...	Aphonia, Dyspnœa, Cough, and Dysphagia.	Nearly whole of Larynx.	Papilloma...
170	1867	M.	60	Druggist ...	Intermittent Aphonia.	Vocal Cords... ...	Two Polypi
171	1867	F.	21	Servant ..	Aphonia and attacks of Suffocation.	Below Vocal Cords...	Sarcoma ...
172	1867	M.	40	Captain	Hoarseness and Oppression in Throat.	Beneath left Vocal Cord.	Polypus ...
173	1867	M.	64		Aphonia since 17 years old; Dyspnœa.	Not stated	Polypus ...
174	1867	M.	34		Attacks of Suffocation.	Posterior Wall of Larynx.	Polypus

Treatment.	Result.	Operator.	Reference and Remarks.
Tracheotomy. Evulsion of Growth with forceps and wire loop.	Cure	Fournié ...	*Gazette des Hôpitaux*, No. 56, 1868.
Destruction with galvanic-cautery loop.	Cure	Idem ...	*Ibid.*
Evulsion with wire loop and application of chromic acid.	Cure ...	Tobold ...	*Berlin. klin. Wochenblatt*, 1869, No. 4. Two or three recurrences ; ultimate cure.
Excision with écraseur ...	Cure	Johnson ...	*Transactions of Medico-Chirurgical Society*, vol. li. Case 7.
Excision with annular knife	Cure	Bruns	*Polypen des Kehlkopfes.* Tübingen, 1868, p. 106.
Excision with knife	Cure	Idem ...	*Ibid.*, p. 110.
Evulsion with wire loop and cauterization.	Improvement	Idem ...	*Ibid.*, p. 114. Tracheotomy had been performed in 1865 ; Canula removed Dec. 15, 1867.
Evulsion with wire loop and destruction with galvanic cautery.	Cure	Idem ...	*Ibid.*, p. 122.
Evulsion with wire loop and cauterization.	Cure	Idem	*Ibid.*, p. 131.
Evulsion with wire loop, destruction with galvanic cautery, and subsequently Tracheotomy.	Improvement	Idem	*Ibid.*, p. 136.
Evulsion with wire loop..	Cure	Idem ...	*Ibid.*, p. 151.
Evulsion with forceps ...	Cure	Dr. Ruppaner ...	*New York Medical Journal*, vol. x. No. 4, p. 337.
Excision with polypotome and destruction with galvanic cautery.	Cure	Voltolini ...	*Die Anwendung der Galvanocaustic, &c.* Wien, 1867, p. 53.
Tracheotomy Thyrotomy, and removal of Growth.	Cure ...	Balassa	*Wiener medizin. Wochenschrift*, No. 92, 1868. Patient returned with recurrence, Feb. 4, 1868. The Thyroid Cartilage was again divided, and the Growth removed *without Tracheotomy.*
Destruction with galvanic-cautery wire loop.	Cure	Voltolini ...	*Berlin. klin. Wochenschrift*, January 20, 1867, p. 27.
Tracheotomy ...	Death ...	Dr. Uhde...	Langenbeck's *Archiv für klinische Chirurgie*, vol. xi. p. 750.
Evulsion with hook forceps, and application of astringents.	Improvement	Schroetter	*Jahresberichte über die Fortschritte.* Virchow, 1867, vol. ii. p. 128.

No. of Case.	Date.	Sex.	Age.	Occupation.	Symptoms.	Situation.	Pathological Nature.
175	1867	M.	56	...	Hoarseness, Shortness of breath.	Both Vocal Cords ...	Sessile Tumour.
176	1867	M.	15	...	Aphonia, Dyspnœa, and attacks of Suffocation.	Left Ventricle ...	Benign (?) Epithelioma.
177	1867	F.	28	... ...	Hoarseness and Cough.	Left Ventricle ...	Uncertain...
178	1868	M.	50	Wholesale Dealer.	Hoarseness ...	Left Vocal Cord ...	Fibroma ...
179	1868	F.	...	...	Hoarseness ...	Whole of right side and one-third of left side of Larynx.	Sarcoma ...
180	1868	M.	...	Clergyman	Hoarseness	Under surface of left Vocal Cord.	Polypus (size of pea).
181	1868	F.	20	Servant ...	Aphonia	Both sides of Larynx	Benign (?) Epithelioma.
182	1868	M.	30	Merchant ...	Hoarseness	Under surface of left Vocal Cord.	Polypus (size of small pea).
183	1868	M.	7		Aphonia	Anterior Commissure of Vocal Cords.	Fasciculated Sarcomatous.
184	1868	F.	...	Girl ...	Aphonia	Left Ary-epiglottic Fold.	Fibroma ...
185	Sept. 4, 1869	M.	38	Wine-Merchant.	Hoarseness, Dyspnœa, and Cough.	Right Ventricle ...	Fibroma ...
186	1869	M.	43	...	Aphonia and Dyspnœa.	Beneath Vocal Cords	Fibroma ...
187	1869	M.	14	...	Aphonia and Dyspnœa.	Anterior Commissure of Vocal Cord.	Papilloma...
188	1870	M.	...	...	Hoarseness and Occasional Aphonia.	Beneath Anterior Commissure of Vocal Cords.	Fibroma ...
189	1870	F.	12	...	Aphonia, Dyspnœa, and Cough.	Whole inner surface of Larynx.	Granular Papilloma.

Treatment.	Result.	Operator.	Reference and Remarks.
Division through Thyro-hyoid Membrane, Thyroid and Cricoid Cartilages ; excision of Growth with curved scissors and cauterization.	Improvement	Dr. Cutter...	*Boston Medical and Surgical Journal,* Feb. 1869. Recurrence on Right Vocal Cord in less than one month.
Tracheotomy, excision with forceps, and cauterization.	Cure	Dr. J. D. Atlée.	*Ibid.,* 1869, p. 262.
Destruction with galvanic cautery and astringents.	Cure	Navratil ...	*Berlin. klin. Wochenschrift,* 1868, No. 49, p. 500.
Destruction with galvanic-cautery porcelain knife.	Improvement	Idem ...	*Ibid.,* p. 500.
Excision with double-edged knife and galvanic cautery.	Improvement	Idem	*Ibid.,* p. 501.
Thyrotomy, and excision of growth with scissors ; subsequent cauterization.	Improvement	Idem ...	*Ibid.,* p. 501.
Thyrotomy and Tracheotomy. Operation abandoned.	Negative ...	Idem ...	*Ibid.,* p. 501.
Thyrotomy, and excision of Growth.	Cure	Idem ...	*Ibid.,* p. 502.
Excision with knife, wire loop, and cauterization.	Improvement	Gottstein ..	*Wiener medizin. Wochenschrift,* Dec. 30, 1868, No. 105.
Evulsion with forceps, scissors, and loop.	Cure	Stoerk ..	*Ibid.,* Nov. 18, 1868.
Tracheotomy, Thyrotomy, and excision of Growth.	Cure	Krishaber ...	*Gazette des Hôpitaux,* 1869, No. 103.
Destruction with galvanic-cautery knife.	Improvement	M. L. Mandl	*L'Union médicale,* 1869, p. 959.
Evulsion with forceps and cauterization	Improvement	Dr. Guyon...	*L'Union médicale,* 1870, p. 705.
Evulsion with forceps and scissors.	Cure	Dr. Giuseppe Gentile ...	*Il Morgagni,* anno xii. Dispensa 6ª, 1870.
Partial excision and cauterization.	Improvement	Idem ...	*Ibid.,* Dispensa 9ª, 1870.

  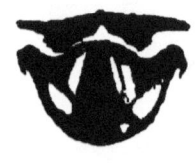

 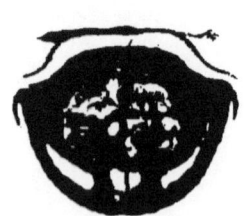 

DESCRIPTION OF PLATE II.

— —

Fig. 1, Case 23.—Papillomata on the Ventricular Bands and on both Vocal Cords.

Fig. 2, Case 18.—Papilloma on the Right Vocal Cord.

Fig. 3, Case 3.—Papillomatous Excrescences on nearly the entire Lining Membrane of Larynx.

Fig. 4, Case 42.—Fibroma on the Right Vocal Cord.

Fig. 5, Case 38.—Papilloma on the Posterior Wall of Larynx.

Fig. 6, Case 25.—Cystic Tumour on the Epiglottis.

Fig. 7, Case 57.—Papilloma on the Left Vocal Cord, and also on the Under Surface of the Epiglottis. Appearance in deep inspiration.

Fig. 8.—The same. Appearance in forced expiration.

Fig. 9.—The same. Rotation of the Growth beneath the Vocal Cord on attempted evulsion.

Fig. 10, Case 52.—Fibro-cellular Growth on the Right Vocal Cord *in situ*, and after removal.

Fig. 11, Case 64.—Papillomata on the Vocal Cords.

Fig. 12, Case 89.—Angeioma in the Right Hyoid Fossa.

33

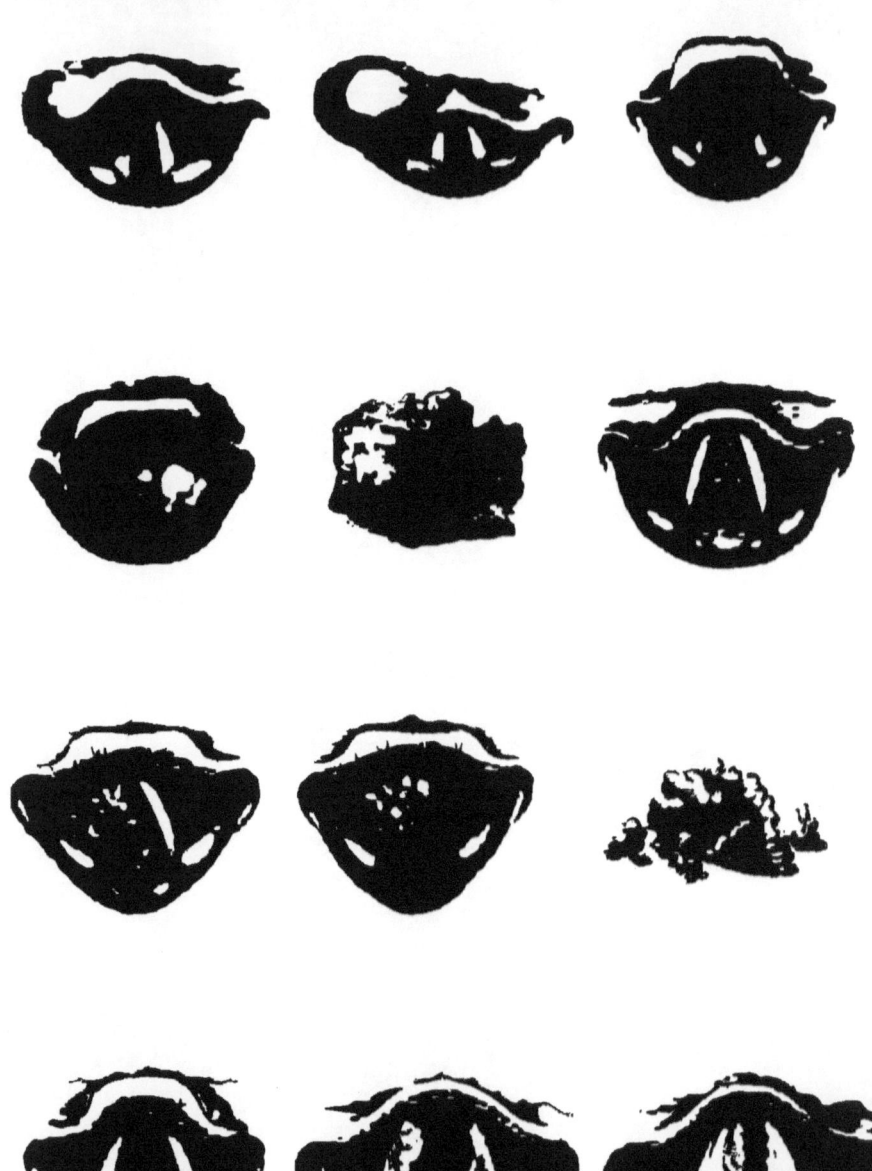

DESCRIPTION OF PLATE III.

—

Figs. 1 and 2, Case 85.—Cystic Tumour on the Epiglottis.

Fig. 3, Case 49.—Fasciculated Sarcoma on the under surface of the Epiglottis.

Fig. 4, Case 88.—Adenoma on the Epiglottis.

Fig. 5.—The Tumour after removal.

Fig. 6, Case 79.—Adenoma beneath the Anterior Commissure of the Vocal Cords.

Fig. 7, Case 95.—Fasciculated Sarcoma on the Right Ventricular Band. Appearance in inspiration.

Fig. 8.—The same. Appearance in Phonation.

Fig. 9.—The same. Tumour after removal.

Fig. 10, Case 97.—Fibroma on the Posterior Wall of the Larynx.

Fig. 11, Case 99.—Fibro-cellular and Myxomatous Growth on both Vocal Cords. The Pink portion on the Right Vocal Cord is Myxomatous.

Fig. 12.—The same twelve days after treatment was commenced.

Plate 4

Fig 2

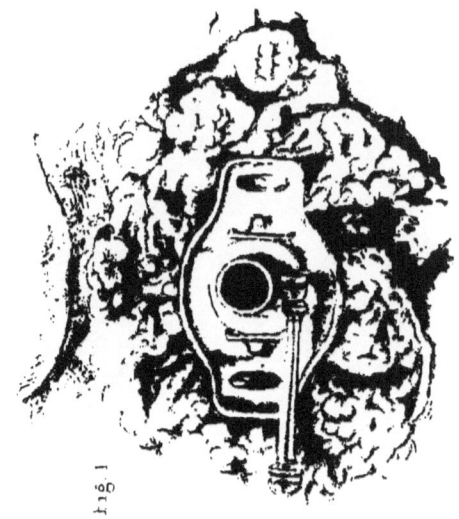

Fig 1

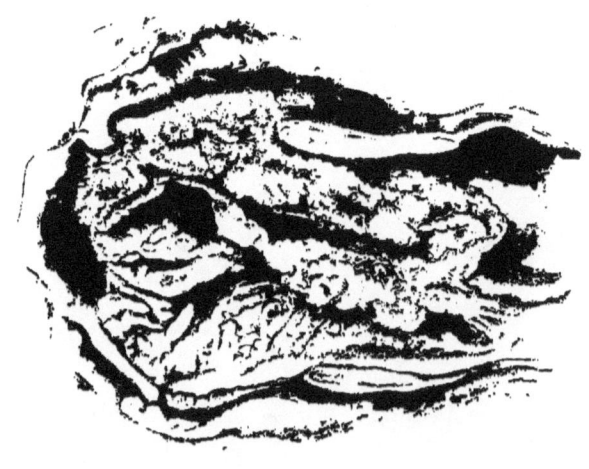

Fig 3

DESCRIPTION OF PLATE IV.

Fig. 1, Case 87.—Epithelioma of the Larynx. External Appearance during life. The canula, which is in the trachea, is surrounded by enormous excrescences growing from the wound and the surrounding skin. (For laryngoscopic appearance, *vide* Woodcuts 74 and 75.)

Fig. 2.—The same. Appearance of Larynx after death, from behind.

Fig. 3.—The same view, but the laryngeal canal has been divided by a perpendicular section.

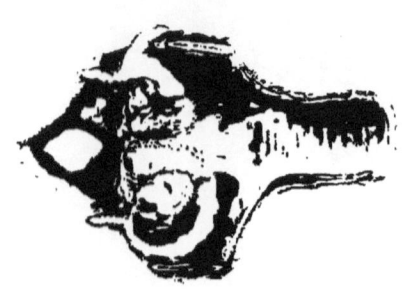

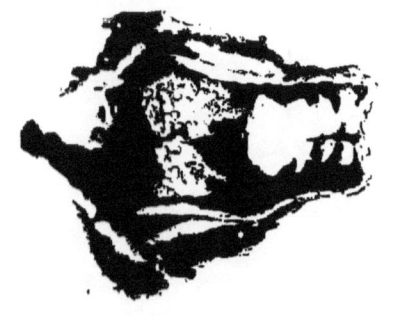

DESCRIPTION OF PLATE V.

Fig. 1, Case 98.—Benign Epithelial Growths on both the Vocal Cords, &c. Appearance after death.

Fig. 2.—Eversion of the Right Ventricle ; Partial Eversion of the Left Ventricle. Appearance after death. (Page 34.)

Fig. 3.—Growths in the Larynx of a dog. Appearance after death. (Page 55.)

INDEX.

34

FINIS.

Wyman & Sons, Printers, Great Queen Street, Lincoln's-Inn Fields, London, W.C.